F. S. Mitchell

The Birds of Lancashire

F. S. Mitchell

The Birds of Lancashire

ISBN/EAN: 9783741124280

Manufactured in Europe, USA, Canada, Australia, Japa

Cover: Foto ©Klaus-Uwe Gerhardt /pixelio.de

Manufactured and distributed by brebook publishing software (www.brebook.com)

F. S. Mitchell

The Birds of Lancashire

THE
BIRDS OF LANCASHIRE

BY

F. S. MITCHELL
MEMBER OF THE BRITISH ORNITHOLOGISTS' UNION

SECOND EDITION

REVISED AND ANNOTATED BY
HOWARD SAUNDERS, F.L.S., F.Z.S., &c.

WITH ADDITIONS BY R. J. HOWARD, M.B.O.U.,
AND OTHER LOCAL AUTHORITIES

ILLUSTRATED BY G. E. LODGE, VICTOR PROUT, &c.

LONDON:
GURNEY & JACKSON, 1, PATERNOSTER ROW.
[*Successors to* MR. VAN VOORST.]
MDCCCXCII.

BOOKS ON BIRDS.

A HISTORY OF BRITISH BIRDS. By the late WM. YARRELL, V.P.I.S., F.Z.S. Fourth Edition, revised to the end of the Second Volume by Professor NEWTON, M.A., F.R.S. The revision continued by HOWARD SAUNDERS, F.L.S., F.Z.S. 4 Vols., 8vo, cloth, with 564 Illustrations, £4.

CATALOGUE OF THE BIRDS OF SUFFOLK, with an Introduction and Remarks on their Distribution. By CHURCHILL BABINGTON, D.D., V.P.R.S.L., F.L.S., &c., with Seven Plates and a Map. 8vo, cloth, 10s. 6d.

THE BIRDS OF MIDDLESEX. A contribution to the Natural History of the County. By J. E. HARTING, F.L.S., F.Z.S. Post 8vo, 7s. 6d.

THE BIRDS OF SOMERSETSHIRE. By (the late) CECIL SMITH, of Lydeard House, near Taunton. Post 8vo, 7s. 6d.

THE BIRDS OF THE HUMBER DISTRICT. By JOHN CORDEAUX, of Great Cotes, Ulceby. Post 8vo, 6s.

THE BIRDS OF NORFOLK. By (the late) HENRY STEVENSON, F.L.S. 3 Vols. 8vo, with Illustrations, £1 11s. 6d.

AN ILLUSTRATED MANUAL OF BRITISH BIRDS. By HOWARD SAUNDERS, F.L.S., F.Z.S., &c. One Volume, 730 pages, demy 8vo, with 367 fine Woodcuts and 3 Maps, £1 1s.

A HANDBOOK OF BRITISH BIRDS. Showing the Distribution of the Resident and Migratory Birds in the British Islands, with an Index to the Records of the Rarer Species. By J. E. HARTING, F.L.S., &c. 8vo, 7s. 6d.

A LIST OF BRITISH BIRDS. Compiled by a Committee of the British Ornithologists' Union. 8vo, sewed, 10s. 6d.

THE BOOK OF DUCK DECOYS, their Construction, Management, and History. By Sir RALPH PAYNE-GALLWEY, Bart. Crown 4to, cloth, 226 pages, with Coloured Plates, Plans, and Woodcuts, £1 5s.

THE FOWLER IN IRELAND : or, Notes on the Haunts and Habits of Wild Fowl and Sea-Fowl, including Instructions in the Art of Shooting and Capturing them. By Sir RALPH PAYNE-GALLWEY, Bart. With Illustrations, 8vo, £1 1s.

FALCONER'S FAVOURITES. By (the late) W. BRODRICK. A series of Life-sized Coloured Portraits of all the British Species of Falcons at present used in Falconry. Large folio, cloth, £2 2s.

FALCONRY IN THE VALLEY OF THE INDUS. By (the late) Sir RICHARD F. BURTON. Post 8vo, with Four Illustrations, 6s.

THE ORNITHOLOGY OF SHAKESPEARE, Critically Examined, Explained, and Illustrated. By J. E. HARTING, F.L.S., F.Z.S. 8vo, 12s. 6d.

HINTS ON SHORE-SHOOTING, including a Chapter on Skinning and Preserving Birds. By J. E. HARTING, F.L.S., F.Z.S. Post 8vo, 3s. 6d.

THE BIRDS OF EGYPT. By G. ERNEST SHELLEY, F.Z.S., F.R.G.S., &c., late Captain Grenadier Guards. Royal 8vo, with Fourteen Coloured Plates, £1 11s. 6d.

A LIST OF THE DIURNAL BIRDS OF PREY, with References and Annotations. By (the late) JOHN HENRY GURNEY. 8vo, sewed, 4s. 6d. ; cloth, 5s.

THE IBIS : A QUARTERLY JOURNAL OF ORNITHOLOGY. Edited by P. L. SCLATER, M.A., F.R.S., &c. 6s. nett. Annual prepaid Subscription, £1 1s.

IN PREPARATION.

THE BIRDS OF IRELAND. By D. M. BARRINGTON, A. G. MORE, R. J. USSHER, and R. WARREN. Demy 8vo.

LONDON :
GURNEY & JACKSON, 1, PATERNOSTER ROW
(SUCCESSORS TO MR. VAN VOORST).

LIST OF ILLUSTRATIONS.

	PAGE
Map of Lancashire	*Frontispiece*
Black-Throated Wheatear	11
Wall-Creeper	60
Martin Mere in 1848	89
Clap- or Cymbal-Nets	93
Plan of ditto	97
Clap-Netter Returning	100
Eagle Crag	133
Plan of Hale Decoy	163
Piel Castle and Donker-Nets	177
Sociable Plover	213
Snipe-Pantle	222
Ring- or Fly-Nets	239

INTRODUCTORY.

This book has been written mainly as a chapter on geographical distribution, a subject which of late years has deservedly received a large share of attention from naturalists, and which when thoroughly worked out for the whole of the British Islands, may be expected to show results both interesting and valuable. It may well be doubted whether the system of taking the counties as limits is not an exceedingly ill-chosen one; but research has in so many instances progressed on these lines, and local enthusiasm is so much more readily stimulated in this direction, that scarcely any choice is left for those portions of the country whose faunal condition yet waits investigation. It can hardly be denied that it would have been far better if, in the division of the ground for local work, regard had been had to physical configuration, and the river valleys, the mountain chains, and the sea-coasts had been taken as boundaries by observers; but the compiler of some future day will have to gather these results for himself, and to collate and compare from this and like histories, whose limits of observation are so arbitrarily defined.

The County of Lancashire is bounded on the west by the sea, on the south by the River Mersey, and on the east, as far north as Pendle Hill, by the chain of hills which here irregularly continues the range of the

Pennines. But further north, the eastern boundary, and also the northern one itself, has no connection with the natural features of the country, and that outlying portion which is inclosed by Windermere, and the Duddon and Winster Valleys, and which forms the district of Furness, ought certainly, for distribution purposes, to be treated in another connection. Throughout the greater part of the county, the character of the land is much the same, and from the sand-hills of the coast, through the mosses adjoining, an almost gradual rise takes place—chequered here and there with moors and "pikes" of no great height—to the high lands of the eastern border. These, while varying from 1,400 to 1,560 feet above the sea-level at the head of Rossendale, rise in Pendle Hill to 1,831 feet; in the moors above the Wyre Valley to nearly as much; and in the extreme north Coniston Old Man reaches 2,653 feet, being surrounded, too, by a group of little less size.

Windermere, Coniston, and Esthwaite are the only inland natural waters of any size, but there are numerous small tarns and meres scattered here and there; whilst the large reservoirs, which supply the towns, form very suitable resting-places for migrating aquatic birds, and the many ponds and small lakes which beautify the residences of the gentry serve in many cases—and might in almost all—as secure breeding retreats for those which remain during the summer.

The rivers are numerous and important, and whilst flowing tumultuously over rocky beds in their upper reaches, in their lower expand mostly into the wide sandy estuaries which are so prominent a feature in the coast line.

The county is very well wooded throughout, but there are no sea-cliffs, and in this respect only are the natural

INTRODUCTORY.

conditions unsuitable for the presence of one or other race of birds.

The vast increase of population, and the scientific farming which drains every marsh and substitutes for every bosky nook a rigid bank and paling mathematically drawn, are the chief causes of the decrease—both in species and individuals—which has taken place in the manufacturing districts; but it is astonishing how many birds still flourish among the teeming millions which dwell there, and should it be possible for air and water to become more pure, there is no doubt that, except in the immediate vicinity of buildings, little further diminution would occur.

The way in which birds are driven away by the extension of buildings, and by the conversion of a rural into an urban locality, may well be instanced by the case of Peel Park, Salford, which is one in point. Mr. John Plant has kindly permitted me to use his notes, which have been carefully kept since 1850, and which show the following results :—

	Personally Observed.	Breeding.
1850–1860	71 species	34 species
1860–1870	42 ,,	
1870–1875	19 ,,	8 ,,
1876–1880	15 ,,	
1881	13 ,,	
1882	5 ,,	2 ,, (Starling & House-Sparrow)

Mr. Plant considers that the main causes are, not so much simply the presence of more people and greater disturbance by them, as the destruction of natural food, and loss of protective foliage from the vitiated atmosphere; and he makes the melancholy prophecy, that, if the same thing goes on for another ten years, there will not be a large tree alive in the park.

The same causes are at work in a less degree in many other districts, but on the whole I do not think that birds generally are decreasing; and the greater scarcity of the Goldfinch, for instance, which feeds on the thistles of waste lands, is balanced by the greater plentifulness of the Hawfinch, which prefers a more cultivated country. Finally, if the game-preserver will lay aside some of his truculence in respect of species which occasionally diminish his stock, if the denizens of towns will discourage the bird-catching fraternity and be content to hear the Linnet and the Bullfinch only in their natural haunts, and if the specimen-hunter will try to be content with skins which are not *local*, there is no reason to expect any approach to extinction of the species which are now on the list.

The avifauna of Lancashire comprises 259 species, which may be sufficiently classed under the following heads :—

Residents ...	Including 85	species.
Summer Visitors	„ 31	„
Winter Visitors	„ 65	„
Occasional Visitors	„ 78*	„
	259	

The residents are all annual breeders within the county limits except the Lesser Black-backed and Herring-Gulls, but as these nest within a very short distance of the border, they may fairly be included in the class. The Peregrine Falcon, Common Buzzard, Hen-Harrier, Nuthatch, Goldfinch, Raven, Rock-Dove, Water-Rail,

* [In 1st Ed., 75. Add, Purple Heron, Sociable Plover, two Petrels, and Black-necked Grebe; deduct Swallow-tailed Kite and Cream-coloured Courser.—Ed.]

and Spotted Crake probably all still breed, though in much diminished numbers. The summer visitors also all breed annually, but the Pied Flycatcher, always local, is now very rare.

Among the winter visitors is placed the Crossbill, which once bred regularly, and possibly still does so occasionally, as are also those species which, like the Dotterel, Greenshank, and Turnstone, appear on migration in spring and autumn, and those like the Guillemot, which occur the summer through, but never remain to breed.

The occasional visitors include the Roseate Tern, which, not many years ago, was a regular summer migrant, and which, though seemingly extinct, I have been reluctant as yet to cut out altogether.

The following species have been procured in a wild state, but most probably have been escapes from confinement :—

Egyptian Goose, *Chenalopex ægyptiacus* (L.); shot several times on the Ribble (J. B. Hodgkinson, R. J. Howard).

Purple Gallinule, *Porphyrio cæruleus* (Vandelli); one shot near Grange, September 25th, 1876 (*Zool.*, 1877. pp. 228, 382, E. T. Baldwin).

Canada Goose, *Bernicla canadensis* (L.); many specimens seen and shot on the coast.

Polish Swan, *Cygnus immutabilis*, Yarrell; has not yet occurred wild, but Yarrell ("Brit. Birds," 3rd edit., 1856) says that at Knowsley a male paired with a Mute Swan, and a brood was produced.*

Mute Swan, *Cygnus olor* (Gmelin); common in a domesticated state on many inland pools and reservoirs.

Lately introduced species of game birds have not succeeded in maintaining a foothold; for instance :

* [A very questionable species.—Ed.]

The Red-legged Partridge, *Caccabis rufa* (L.), was turned down at Rufford in some numbers by the late Sir Thomas Hesketh, about thirty years ago, and by the present baronet in 1879, but all have disappeared.

The Virginian Colin, *Ortyx virginianus* (L.), was introduced by Lord Lilford on the Bank Hall estate, and 500 pairs turned out in the latter part of December, 1874. Gradually the birds have disappeared, and now there is scarcely one left. A good deal of snow was on the ground at the time, but the birds seemed to do well until the breeding-season, when their clear whistle might be heard all over the moss. They were very jealous of any interference with their nests, many of which were found forsaken. Mr. R. J. Howard, who has procured me this information, says that it is supposed the disturbance caused by farm labourers and others crossing the country prevented their thriving; and on account of their pugnacity, they are also credited with the present scarcity of the Common Quail in that locality.

First and foremost of species which have been recorded as having occurred in Lancashire, as I think, on insufficient grounds, comes the Nightingale *Daulias, luscinia* (L.). In Cheshire, I believe, there is no doubt of its occasional presence, but in Lancashire I do not find any evidence worth a moment's attention, except that provided by Mr. R. Standen, of Goosnargh, in the *Field Naturalist and Scientific Student*, part 2, Manchester, 1882–83. This gentleman therein states, and has amplified the matter in correspondence with myself, that in 1864 or 1865, in June, he heard the Nightingale singing at Samlesbury, and in June, 1871, at Whittingham Hall. On the latter occasion he says that he *saw* the bird twice, and that it sang for nearly a fortnight, when it was probably driven away by attempts of village lads to capture it. In the former instance, a friend from Hampshire, familiar with the song, heard it, and was convinced it was the Nightingale. This would seem fairly conclusive, but every experienced field-naturalist knows how easy it is to mistake a bird when seen dimly among thick foliage; and evidence from the hearing of the song can only be accepted when the observer is not only familiar with the particular notes, but is also possessed of an ear trained by long use to distinguish the—often very similar—songs of allied species. It would, doubtless, not have been difficult for Mr. Standen to settle the question by shooting either of these songsters; but I think he is to be congratulated for having refrained, and that it is more to his honour that room is left for scepticism, than that he should be able to point to a skin in his cabinet as a specimen of what might have been the progenitor of a race of Lancashire Nightingales.

Other insufficiently verified species are:—

Water Pipit, *Anthus spipoletta* (L.); Mr. T. Webster, of Manchester, writes (*Zool.*, p. 1,023, June 11th, 1845) that he saw, in October, 1843, at Fleetwood, three birds which he identified as of this species from reading a diagnosis of *Anthus aquaticus*, one of the synonyms of *A. spipoletta*.

Tawny Pipit, *Anthus campestris* (L.); Mr. John Hardy, of Manchester, writes to me that he "has seen one or two skins of the Tawny Pipit, and apparently in a fresh condition, which were *said* to have been shot near the Bolton reservoir at Entwistle."

Dartford Warbler, *Melizophilus undatus* (Boddaert); Mr. T. Webster (*Zool.*, 1845, p. 1190) believes he saw a pair with three or four young, at Lytham, about August 27th. 1845.

Red-breasted Goose, *Bernicla ruficollis* (Pallas); Mr. Hugh P. Hornby writes November, 1882, and with reference to a note of his in the *Zoologist* for 1872, p. 3236, "I quite believe that the two geese shot over twenty-five years ago on Sowerby meadows (near St. Michael's-on-Wyre), and called by our keeper, R. Crook, 'Siberian Geese,' belonged to this species."

Owing to a misprint for Lincolnshire in Montagu's "Ornithological Dictionary," 1st edit., 1802, the Bearded Tit, *Panurus biarmicus* (L.), has been erroneously recorded by many writers as a Lancashire species, and in *The Zoologist*, 1879, p. 305, I pointed this out.

The Stone-Curlew, *Œdicnemus scolopax* (S. G. Gmelin), has also been catalogued as a Lancashire bird ("Fauna of Liverpool," 1856, Byerley; *Nat. Scrap Book*, part 12, R. Reynolds), but in all cases I believe there has been an error in identification.

The dates of arrival and departure of migrating species can only be given approximately, and there is no doubt that these vary considerably with the character of the season; that is, in relation to the food-supply, and not so much to a few degrees of temperature, for it is warmer when birds go than when they come. The observations of the late Mr. Blackwall ("Researches in Zoology," 1834, p. 8 *et seq.*) have shown that, taking a hap-hazard number of years, 1817 to 1828, the mean temperature of September and October was higher than that of April, and—in the cases of the Cuckoo and the Swift especially—that which prevailed at their departure was uniformly higher than on their arrival.

The uselessness of averages may be seen by a glance at the annexed tables compiled by Mr. Blackwall, the first of which gives the average dates of arrival round Manchester of a series of summer visitors from 1814 to

1821 ("Mem. Manchester Lit. and. Phil. Soc.," 1824), and the second from 1814 to 1828 ("Researches in Zoology").

Birds.	Eight years. From 1814 to 1821.	Fifteen years. From 1814 to 1828.
Sand-Martin	Appear April 6	Appear April 9
Wryneck	,, ,, 6	,, ,, 11
Willow-Wren	,, ,, 12	,, ,, 11
Wheatear	,, ,, 14	,, ,, 6
Swallow	,, ,, 18	,, ,, 15
Whinchat	,, ,, 21	,, ,, 26
Blackcap	,, ,, 22	,, ,, 27
Martin	,, ,, 23	,, ,, 25
Cuckoo	,, ,, 24	,, ,, 20
Grasshopper-Warbler	,, ,, 30	May 5
Whitethroat	,, May 2	,, April 30
Spotted Flycatcher	,, ,, 14	,, May 11
Sedge-Warbler	,, ,, 19	,, ,, 3

The fact is, that very few of the observations, now so numerously made, as to the movements of summer migrants, are worth anything at all; and if data are to be collected on land of a value commensurate with those now being collated on information from lighthouses, &c., by the Committee appointed by the British Association, it will be necessary for the observer to fulfil something like the following conditions. Firstly, that he should be continuously engaged out of doors; secondly, that he should be entirely familiar, not only with the plumage of birds, but that he should be able to recognize most of them when flying, and be thoroughly acquainted with their song, their call and alarmnotes; thirdly, that he should have a knowledge of the food requirements of each species, and be able, for instance, to infer, from the plentifulness of such and such an insect, that such and such a bird may be expected to feed on it. Such a conjunction can only be found in few individuals; but if every man in his leisure field-walks would—especially in connection with meteorological conditions—note the other natural circumstances at the time of his first seeing a spring arrival, a mass of information would be got together which might be invaluable for the discovery of the laws of geographical distribution. Until something of the sort is done, and such information sifted and compared, I believe those laws will remain, as they are now, dubious and conjectural.

The winter migrants, or more especially those species which visit our shores periodically in autumn and spring, are easier of observation, and the British Association Committee above mentioned have already been able to indicate some probable results as to their lines of flight and the causes of their movements.

Although few of the Geese, Ducks, Tringæ, &c., which breed in the north, fail to visit the Lancashire coast, on their passage to and fro, in small numbers, the large flocks which are seen on the coasts of Yorkshire, Norfolk, and the east generally, are not found on the west, and the streams of migrants there are mostly composed of species which breed in Scotland, Iceland, and Greenland, Scandinavian and Siberian forms being very irregularly represented.

Results from Lancashire lighthouses, too, are small, and indeed seem to be fewer now than formerly. Mr. W. A. Durnford (*Zool.*, 1876), remarking on a Kingfisher which was killed by flying against that on the south end of Walney Island, says that the light-keeper told him that thirty years before it was not unusual for 100 birds to kill themselves against the glass in a single night, whilst during the previous six months a Stock-Dove and a Woodcock had been the only victims. Knots and Grey Plovers, however, occasionally immolate themselves, as also a few Starlings, Blackbirds, Thrushes, Cuckoos, and Curlews.

Of the literature of the subject there is little to be said. No attempt that I know of has been made to focus the condition of the county as regards ornithology beyond the " Catalogue of Birds found in Lancashire," compiled by Mr. Peter Rylands and published in the *Naturalist* for 1837, and the papers on the " Birds of Lancashire and Cheshire," written by Mr. Frank Nicholson in the *Manchester City News* of 1875, and both these are necessarily very much curtailed. " The Natural History of Lancashire, Cheshire, and the Peak in Derbyshire," by Dr. Charles Leigh, and published at Oxford in 1700, contains many records of that period which are exceedingly interesting, and, although disfigured by gross absurdities, I see no reason to doubt the correctness of the ordinary information. The good Doctor avers that " these counties afford us great variety of birds, and in some places even clog the inhabitants with their plenty," and is always very careful to state the capabilities of each species from a culinary point of view. There are " no counties in England," he says, " affording so great a variety of mines, minerals, and mettals, with other choice products, and the most surprising phænomena of nature," and takes the opportunity of introducing among his " phænomena " the woman at Whalley who had two horns growing out of the back of her head. From the " Ornithology of Francis Willughby, F.R.S.," by John Ray, F.R.S., London, 1678, I have culled some valuable notes; and in the " Journal of Nicholas Assheton of Downham," from May 2, 1617, to March 13, 1618 (ed. Raines,

Chetham Soc., 1848), the jovial squire relates how that, on November 24, he "had some sport at Moorgame with my piece, but killed not," and that on December 23, he went "to Rowe Moore, and killed ther three heath cocks."

The following list of papers, periodicals, and books, which I have examined, will show the main sources from which I have derived my published information: Camden's "Britannia," translated from ed. of 1607, and enlarged by Richard Gough, 2nd edit., 1806; "Antiquities of Furness," T. West, London, 1774; "British Zoology," Thos. Pennant, Warrington, 1776-77; "Harmonia Ruralis," James Bolton, 1794; "Nat. Hist. of British Birds," E. Donovan, London, 1794-1819; "A Tour from Downing to Alston Moor," Thos. Pennant, 1801; "Rural Sports," Rev. W. B. Daniel, London, 1801-13; "Beauties of England and Wales," James Britton, London, 1807; "A Companion to the Liverpool Museum," Wm. Bullock, 6th ed., 1808, 7th ed., 1809; "British Ornithology," Geo. Graves, London, 1821; "A General History of Birds," John Latham, M.D., Winchester, 1821; "Proceedings of the Manchester Literary and Philosophical Society," 1785, *et seq.*; "Proceedings of the Liverpool Literary and Philosophical Society, 1812, *et seq.*; *Zoological Journal*, 1824-34; "History of Lancashire," John Corry, 1825; "History of British Birds," Thos. Bewick, 1832; *Field Naturalist*, ed. J. Rennie, 1833-34; "Researches in Zoology," John Blackwall, 1834; *Magazine of Natural History, &c.*, 1828-40; *Annals of Natural History*, 1838-40; *Annals and Magazine of Natural History*, 1840, *et seq.*; *The Naturalist*, 1836-39, 1851-57, 1864-67, 1875, *et seq.*; "History of British Birds," W. Macgillivray, 1837; "History and Antiquities of the Abbey of Furness," T. A. Beck, 1844; "Proceedings and Papers of the Historic Society of Lancashire and Cheshire," 1848, *et seq.*; "Natural History of Selborne," Rev. Gilbert White, ed. Jesse, 1851; "Eggs of British Birds," Hewitson, 3rd ed., 1856; "British Birds," Yarrell, 3rd ed., 1856, and 4th ed., 1871-85; "The Fauna of Liverpool," Isaac Byerley, 1856; *Zoologist; Field;* "Reports of Liverpool Naturalists' Field Club," 1860, *et seq.*; *Ibis;* "Reminiscences of an Old Sportsman," Col. J. P. Hamilton, 1860; *Naturalists' Scrap Book*, 1863-64; *Science Gossip;* "Montagu's Dictionary," ed. Newman, 1866; *Land and Water; Liverpool Naturalists' Journal*, 1866, *et seq.*; "Papers, Letters, and Journals of William Pearson, of Crosthwaite," 1863; "Occurrences of Grey Phalarope in 1866," J. H. Gurney, Jun., 1867; "Portfolio of Fragments," Gregson, 1869; "Report of Bury Natural History Society," 1871; "Handbook of British Birds," J. E. Harting, 1872; "History of Whalley," Whitaker, 4th ed., 1872-76; *Yorkshire Naturalists' Recorder;* "A Cavalier's Note Book," ed. Rev. T. E. Gibson, N.D.; "Lancashire and Cheshire, Past and Present," Thos. Baines, N.D.; "Manchester Walks," Grindon, 1882; *Field Naturalist*, Manchester, 1882-83; "List of Birds found in the Neighbourhood of Walney Island," W. A. Durnford, 1883; "Birds of Europe," H. E. Dresser, 1871-83.

Many scattered passages are found in the pages of old writers showing the then wild and uncultivated state of the country; and the wide expanses of land—now bare and comparatively treeless—which go by the names of the Forests of Bowland, of Pendle, of Trawden, and so on, indicate the vast extent of the changes which have come about in modern times. Shortly before the beginning of the present century the last of the Red Deer, which once roamed through Bowland Forest, were destroyed, and at Brosholme Hall, the residence of the Parkers (hereditary bow-bearers) is still kept an iron stirrup or ring, through which all dogs permitted to live had to pass, in order to ensure the safety of the deer.

It was very common in old grants for hawks (Peregrine Falcons, no doubt) to be reserved with the animals preserved as game; and Beck ("Hist. and Antiq. of Furness Abbey," 1844) states that, at the second crusade, Robert de Boyville mortgaged land at Kirksanton and Horrum, and his cousin Henry confirmed the mortgage, but reserved to himself buck, doe, wild boar and an *aery of hawks*. By some ancient grants also recited by West ("Antiq. of Furness," 1774), it appears that wolves, wild boars, deer, falcons, &c., were common in that district, and that Richard de Lucy, Earl of Egremont, who was Lord Chief Justice of England in the reign of Henry the Second, in a grant to Reginald Fitz-Adam, makes this reservation: " Salcis mihi et heredibus meis cervo et cerva, apro et leia, et *accipitre*, quando ibi fuerint." Leland, who wrote in the sixteenth century, says of Salfordshire ("Itinerary," 1770, vol. vii. p. 49) that, " wild Bores, Bulles, and Falcons bredde in times paste in Blakele ; " and the pastime of hawking was no doubt a favourite one, footpaths, called *hawk-paths*, being formed among the mosses of the Fylde, from which the sport might be more easily followed ("Lanc. and Chesh., Past and Present," Thos. Baines).

Whatever remarks have been made on the habits of the various species are the result of independent observation, and in all cases have a local bearing. It would have been easy to have doubled the size of the book by giving full descriptions of nests and plumages, and by copying largely on such points from the many excellent works now published, but no single fact that I know of proceeds from any but an original source. The writings of Lancashire men in the various periodicals have, of course, been utilized, as these come naturally within the scope of the subject. Where no authority has been given for a statement, it may be taken as proceeding from the personal experiences of myself and my constant out-of-door companion, Mr. Thomas Altham, of Clitheroe, one of that race of artisan-naturalists of which, though more especially in regard to Botany, Lancashire has such reason to be proud.

The time at about which each species lays its eggs has been given as near as experience has taught to be correct, but birds not naturally double-brooded will, if any accident has happened to their first nest, lay a second clutch a few weeks later on. I do not know that any, except those of which it is so indicated, naturally bring up two or more broods in a season, and in the case of late nests, to add to the list, it would be necessary to watch special pairs, and see whether their first attempt had been successful or not.

I gratefully acknowledge the help I have received from naturalists in all parts of the county, and from the list which follows, showing the boundaries which limit the observations of each, it will be seen that, taken in conjunction with published matter, the whole of the ground has been thoroughly worked:—

Mr. C. E. READE, Manchester.
    Notes from Urmston have their limit about eight miles from Manchester, and on the south-west side only. Notes from Moston, three to five miles north of Manchester.

Mr. JOHN HARDY, Manchester.
: South and south-east of Manchester, and immediately contiguous to the city itself.

Mr. J. CLAYTON CHORLTON, Withington.
: Boundaries are Stockport, Stretford, north bank of Mersey, and Rusholme; comprise Chorlton-cum-Hardy, Withington, Didsbury, Burnage, Heaton Mersey, and Heaton Chapel.

Mr. HENRY KERR, Stacksteads.
: Bounded on north and west by Hameldon and Cribden, on the south by Cowpe Law and Brandwood Moor, and on the east by Tooter Hill, Thieveley Pike, and Deerplay Hill.

Mr. R. J. HOWARD, Blackburn.
: A line drawn through Colne, Wigan, Southport, Preston, and Chipping; the neighbourhood of Hawkshead, the Lancashire portion of the Lake District, and that part of the count about Leck.

Mr. R. STANDEN, Goosnargh.
: Bounded on the south by Grimsargh, Haighton, Preston, and Ashton-on-Ribble; on the west by Cottam, Woodplumpton, Broughton, Barton, and Claughton; on the north by the Brock (Brock bottoms), Bleasdale, Fair Snape and Saddle Fells; and on the east by Thornley and Longridge Fells, Ribchester and Alston.

Mr. J. B. HODGKINSON, Preston.
: Neighbourhood of Preston and Witherslack.

Mr. J. A. JACKSON, Garstang.
: Within radius of four miles from Garstang station.

Messrs. HUGH P. HORNBY and JAMES J. HORNBY.
: Neighbourhood of St. Michael's-on-Wyre.

Mr. JOHN WELD [the late], Leagram.
: Within five or six miles from Leagram Hall.

Mr. RICHARD RAMWELL, Turton.
: Neighbourhood of Turton.

Mr. RICHARD DAVENPORT, Holcombe.
: Neighbourhood of Bury and Bolton.

Mr. HENRY MILLER, Accrington.
: Neighbourhood of Accrington and Whalley.

Mr. THOMAS JACKSON, Overton.
: Most of the country between the river Lune west of Lancaster and Morecambe Bay, as also the Lune estuary.

Dr. C. A. PARKER, Gosforth.
: The Cumberland border.

Mr. W. A. Durnford's "Birds of Walney" includes the country within a radius of thirty or forty miles from Barrow-in-Furness; the notes taken by the late Mr. W. Pearson refer to the valley between Underbarrow Scar and Cartmel Fell, and include Whitbarrow, and Lythe Marsh at head of Morecambe Bay; the 1871 "Report of the Bury Nat. Hist. Soc." takes in almost the whole of East Lancashire, and the limits assigned to the contributors to the *Naturalists' Scrap Book* were a circuit of fifteen miles round Liverpool, with two miles round Southport added, but the Flintshire coast excluded.

In addition to the above, I have received much valuable information from the following gentlemen, who, by the trouble they have taken in aiding me to clear up many doubtful points, have placed me under great obligations:—

Rev. E. D. Banister, Whitechapel, Preston;
Mr. W. E. Beckwith, Eaton Constantine, Salop;
 ,, W. Fitzherbert Brockholes, Claughton;
 ,, Arthur Breakell, Garstang;
 ,, R. Drummond, Blackpool;
 ,, John Drake, Bury;
 ,, James Garnett, Clitheroe;
 ,, W. Gillett, Preston;
 ,, John Hall, Nateby;
 ,, John Hancock [the late], Newcastle-on-Tyne;
 ,, J. E. Harting, London;
 ,, James Holland [the late], Middleton;
 ,, Thomas Jones, Church;
 ,, Wright Johnson, Prestwich;
 ,, Rawdon B. Lee, Kendal;
Lord Lilford, Oundle;
Mr. Anthony Mason [the late], Grange;
 ,, T. H. Miller, Singleton;
 ,, T. J. Moore, Liverpool;
 ,, F. Nicholson, Altrincham;
 ,, J. E. Palmer, Dublin;

Mr. W. PETERKIN, Clitheroe;
,, JOHN PLANT, Salford;
,, HOWARD SAUNDERS, London;
,, LOUIS H. SIMPSON, Preston;
,, J. P. CHAMBERLAIN STARKIE [the late], Ashton Hall, Lancaster;
,, J. A. ST. CLAIR, Blackpool;
,, JOHN SUMNER, Halsall;
,, R. H. THOMPSON, Lytham;
,, JOHN WATSON, Kendal;
,, HENRY WHALLEY, Colne;
,, J. H. WOOD, Bury;
,, JOHN WRIGLEY, Formby.

The classification adopted is [mainly] that of the List of British Birds compiled by a committee of the British Ornithologists' Union, published in 1883.

F. S. M.

CLITHEROE,
December, 1884.

PREFACE TO THE SECOND EDITION.

THE First Edition having been exhausted very soon after its appearance in 1885, and several species having since been added to the Lancashire list, I accepted the task (in the absence of Mr. Mitchell from England) of preparing a Second Edition, as far as practicable, up to date. I should have preferred that this had devolved upon some ornithologist who was personally connected with the county, but my deficiencies in this respect have, I trust, been remedied by the cordial assistance of Mr. R. J. Howard of Blackburn, whose name appears on the title-page, and whose valuable notes will be found throughout the work. To him also we owe some recent details from Messrs. W. Fitzherbert Brockholes, Hugh P. Hornby, and others. Again, my thanks are due to Messrs. Frank Nicholson and C. F. Archibald; and especially to the Rev. H. A. Macpherson, of Carlisle, who generously placed at my disposal the proof-sheets of his "Fauna of Lakeland," in which there is some important information respecting the outlying district of Furness. My own share has been chiefly editorial, and in this connection my labours have been lightened by the admirable and systematic Bibliography for the Northern Counties, published in the present series of *The Naturalist*. The species added to the Lancashire list in this edition are the Purple Heron (p. 145), Sociable Plover (p. 213, with woodcut), White-faced Petrel (p. 258), Wilson's Petrel (p. 258), and Black-necked Grebe (p. 262). There is a new Index. Inasmuch as the full-page illustrations of Duck-decoys in the first edition relate to Fritton in Suffolk, and have, moreover, already appeared elsewhere, they are now omitted.

H. S.

7, RADNOR PLACE, HYDE PARK, W.,
    LONDON,
        August 1st, 1892.

THE
BIRDS OF LANCASHIRE.

ORDER PASSERES.

FAMILY TURDIDÆ.—SUBFAMILY TURDINÆ.
GENUS TURDUS.

MISTLE-THRUSH.

TURDUS VISCIVORUS, Linnæus.

LOCAL NAMES—*Sher-cock, Chir-cock, Set-cock, Shirley, Shirley-cock, Storm-cock, Shrite-cock, Swine-throstle.*

A resident species, breeding more or less numerously throughout the county. In autumn it congregates in small flocks, so remaining all winter, at which season it is very wary. During the breeding-season also it is more retiring than its congeners, being much more often found in the woods of the fells than they are; occasionally, however, it breeds near houses, possibly seeking protection from Rooks and Magpies. It usually lays four eggs, but sometimes five, and Mr. J. P. Thomasson says (*Zool.*, 1861) that he has known it in several instances to sit on three eggs only; although when disturbed at this time it is generally very noisy, sometimes it will fly away quite silently. The nest, usually lined with dry grass, but sometimes plastered a little with mud like that of the Song-Thrush, is often placed high up in the trees, and would be difficult to see, were it not that pieces of sheep's-wool, half loose, almost always depend from it in very slovenly fashion. The Mistle-Thrush is double-brooded, and the first lot of eggs is laid from the end of March to May, but most generally in April. It appears to increase in numbers, and Mr. Hugh P. Hornby says that at St. Michael's-on-Wyre it is most plentiful when hard weather begins to set in, remaining so long as there are any yew-berries to feed on. Mr. J. Plant, writing in 1876, says that it only visits Peel Park, Salford (coming to feed on the hawthorn-berries) in severe winters, but when these are protracted, Mr. J. Hardy says it disappears altogether from the neighbourhood of Manchester. The hard winters of 1878, 1879, and 1880 thinned its numbers very much. Its song, which is louder than that of the Song-Thrush, though not so varied, may be heard the year through; except in the severest weather, when it only utters its whistling call-note.

[Mr. R. J. Howard says that on the edge of the fells the Mistle-Thrush nests on the tops of dry stone walls. On May 14th, 1892, he visited a nest from which Mr. Altham had taken two eggs on 7th, substituting two of Blackbird; by 14th the Mistle-Thrush had laid three more eggs, and had just hatched one of the Blackbird's. The nest was on the ground at the edge of a clough: a situation in which Mr. Howard had previously found nests of the Song-Thrush and Blackbird, but never one of the Mistle-Thrush.—Ed.

SONG-THRUSH.

TURDUS MUSICUS, Linnæus.

LOCAL NAMES—*Throstle*, *Mavis* (rarely, in Furness).

Resident and common everywhere, even in the neighbourhood of the largest towns, and Mr. J. Hardy informs me that near Manchester it breeds regularly, chiefly in gardens and enclosures. In severe winters it is partially migratory, many birds going southwards, and Mr. John Weld, of Leagram Hall, says that it retires altogether from the vicinity of Chipping. Its plastered nest is well known, and the late Mr. Thomas Garnett, of Clitheroe, wrote (*Mag. of Nat. Hist.*, 1830): " I agree with Mr. Jennings that the Throstle does not line its nest with mud, but generally with some compost of which cow-dung forms a part, although I have found [nests] lined entirely of rotten wood. It is a fact also that it invariably lays the first egg whilst the lining is wet." The Rev. Henry Berry says (*Mag. of Nat. Hist.*, 1834): "I have known Throstles, which had been robbed of their nests after one or two eggs had been laid, rebuild in a surprisingly short period and even upon the old foundation. I once took a nest, containing three eggs, but accidentally left behind the coarse external part of the nest; circumstances led me by the place on the following morning, when I observed the Throstle seated on the remnant of her nest, in which she had deposited her fourth egg, having, since the day before (that of the robbery), plastered it with the *usual* coating of *rotten* wood, moist earth, and perhaps a little cow-dung." The same gentleman also says (*Mag. Nat.*

Hist., vol. vii. p. 598): "With respect to the Thrush, I recollect a singular case: in the garden of James Hawkin, a nursery-man at Ormskirk in Lancashire, a Thrush and Blackbird had paired: this was well known to a number of individuals, myself among them. During two successive years, the birds reared their broods, which were permitted to fly, and evinced, in all respects, the features of strongly-marked hybrids." In fine weather the Song-Thrush will sing in every month in the year, and Mr. T. Altham, of Clitheroe, thinks that young birds of the year sing quietly in autumn. On the question of autumn-singing, Mr. John Blackwell remarks ("Researches in Zoology," p. 53), "several species of birds which cease singing about the latter end of July, or the beginning of August, are sometimes heard again in autumn; when their songs are generally feeble, imperfect, and of short continuance, like the early efforts of our warblers in spring." The Song-Thrush lays four or five eggs and hatches two broods in the season. April and May are the great breeding months, but eggs are often taken in March, and in Ribblesdale these are supposed by the natives to be laid by a different species, which they call the March Throstle. A correspondent writing to the *Field* of January 29th, 1859, from Liverpool, says, under date January 26th: "As a proof of the mildness of the season . . a Thrush has a nest in my garden, with one egg, on which she is sitting. During the winter, the Thrushes have sung, and continue to do so almost every day." Terrible havoc was made among them by the winters of 1878, 1879, and 1880 (in that of 1878 they left the Ribble valley altogether); and Mr. C. S. Gregson says (*Zool.*, 1879) that, on his warren near Formby, in December, Song-Thrushes were in "hundreds—aye, thousands," no doubt migrants from the more northern districts. In the springs of 1883 and 1884, however, they appeared to have recovered their normal numbers.

REDWING.

Turdus iliacus, Linnæus.

A winter visitor, appearing usually the latter half of October, but in mild seasons in the earlier half; if very severe weather occurs on its arrival, it goes south at once. In most parts of the county it is found in considerable numbers; but, according to Mr. H. Kerr, it is only occasionally seen in Rossendale, and Mr. T. Jackson has observed very few in the neighbourhood of Overton, on the Lune estuary. Mr. J. Hardy says that it is much less abundant than formerly near Manchester, and in the Urmston district also Mr. C. E. Reade states that its numbers have diminished. At St. Michael's-on-Wyre, Mr. Hugh P. Hornby thinks that the Redwing suffered more than any other of the family during the hard winters before mentioned, and considers that, whilst before a common winter visitor, it is now a rare one. It leaves again for its northern breeding-haunts in March or early April.

[Mr. R. J. Howard informs me that great numbers feed under and roost in the rhododendrons at Billinge Hill, and even during the severe winter up to the end of March 1892, hundreds were about in the valleys and wooded districts.—Ed.]

FIELDFARE.

Turdus pilaris, Linnæus.

Local Names—*Feljar, Blue-back.*

A winter visitor, remaining from October to May, though seldom seen later than April anywhere but in the sheltered district around Grange on the north side of Morecambe Bay. It is usually a fortnight later both in its arrival and departure than the Redwing. The late Mr. W. Pearson, of Crosthwaite, in some notes read before the Kendal Natural History Society, on December 8th, 1839, referring to the Winster valley, remarked: "Fieldfares and Redwings do not now stay with us through the winter. They come in large flocks about the latter end of October, and we see them again in spring. In my youth we used to have them the winter through, feeding on holly-berries and haws." In all other parts of the county, however, the Fieldfare appears to remain the winter through, although near Manchester it is every year becoming less numerous, formerly being moderately abundant. Mr. T. Jackson also says that in his district of Overton it is decreasing; there it is to be seen on the meadows feeding among the Lapwings, but when the ground is covered with snow, it takes to the hedges and feeds on the haws. It has never been known to remain to breed, though the late Mr. J. F. Brockholes reported (*Proc. Liverpool Lit. and Phil. Soc.*, 1859-60) having seen one on an unfinished nest at Maghull.

[Mr. R. J. Howard says that the Fieldfare frequents higher and more open ground than the Redwing, except during the hardest weather; and this is decidedly my own experience.—Ed.

BLACKBIRD.

Turdus merula, Linnæus.

Local Names—*Ouzel, Black Ouzel.*

A resident and common species, and able to stand hard winters better than others of its family. Hence, since 1880, it has been more numerous than the Song-Thrush, whilst usually less so. Like the latter species it is still moderately common on the south side of Manchester, chiefly in gardens and enclosures, and breeds regularly, bringing up two broods in the season. It sings from March to the end of July, and is one of the earliest risers in the morning; its call, in the height of summer, may always be heard before sunrise. Its nest is not seldom placed on the ground, and a case I reported in the *Zoologist* for 1877 showed extraordinary perseverance under difficulties. This nest was in the bottom of an old lime-quarry, placed on a sloping bank, with too little solid foundation, and the materials kept slipping down the bank of their own weight, till a queue nearly two feet long and five inches wide was made. At the head of this it was at length triumphantly completed, and the eggs laid in due course. May is the usual breeding month, though many nests may be found in April and June. The Blackbird probably lays eggs abnormal in shape more frequently than any other common bird; very often they are nearly round, and one in my collection, out of a nestful of the same shape, is only one-tenth of an inch greater in length than breadth. The number is almost always four.

RING-OUZEL.

TURDUS TORQUATUS, Linnæus.

LOCAL NAMES—*Rock-Ouzel*, *Fell-Ouzel*.

A summer migrant, arriving from early in March to early in April, and often remaining till October, though most leave in September. It is very generally distributed over almost all the higher lands, seldom nesting on the lower levels, though Mr. John Hardy says that it breeds occasionally on the drier parts of Chat Moss, Barton Moss, and surrounding neighbourhood, and Mr. C. S. Gregson (*Naturalist's Scrap Book*, 1863–64, pt. 8) records that he has taken its nests on the banks of the Irwell, Irk, Medlock, and Mersey, the whole within a few miles of the city of Manchester. It breeds in Bircle, an elevated district near Bury, and is pretty plentiful on the whole range of moors on the Yorkshire border, from Blackstone Edge northwards to the Leck Fells, preferring those that are bare and rocky, and intersected with old walls. It is less common on Holcombe Hill, but Mr. H. Miller has not seldom found its nests in the clefts on Haslingden Moor and Hapton Scouts. Col. H. W. Feilden says (Dresser's "Birds of Europe") that it is common during the breeding-season on Withnell and Anglezark Moors, between Bolton and Chorley, and that there he has invariably found the nests in banks of water-courses. On Pendle Hill there are always a few pairs, and they breed among a lot of furze-bushes at the bottom, where also are quantities of Blackbirds, nests taken here requiring careful identification. But the Ring-Ouzel is not often found as low down the hills as the Blackbird, and with an occasional Mistle-Thrush (here nesting on the hill-walls), is the only one of its genus that breeds at any altitude. In the Furness district Mr. W. A. Durnford says ("Birds of Walney," 1883) that it nests amongst the hills of the mainland, but is not common, and Mr. John Watson also thinks it is getting rarer, though often met with in the end of March, on the lower lands, before it flies to its breeding-haunts on the fells. The nest—usually well concealed—is very like that of a Blackbird in construction, though rather shallower, and is placed under furze-bushes, among heather near turf pits, or in a cavity of an old wall or rocky bank, near a stream if possible. The eggs are laid from the beginning of April to the beginning of June, and are almost invariably four in number. In its habits the Ring-Ouzel is very wild and shy.

GENUS SAXICOLA.

WHEATEAR.

SAXICOLA ŒNANTHE (Linnæus).

LOCAL NAMES—*White rump, Whit-tail, Wall-tack, Wall-check, Stone-check, Stone-smack, Stone-smatch, Clod-hopper.*

A regular summer visitor, though local in its distribution. The first arrivals appear about the last week in March, the main body in April, and the middle or end of September is the average time of departure. The Wheatear is found breeding on the whole range of coast sand-hills from Liverpool northwards, but elsewhere is most common on the moors and fells, not frequenting

much the lower lands, except where large tracts of
"moss" form a solitude which suits its retiring habits.
It is particularly numerous on the island of Walney, and
Mr. J. E. Harting says (*Zool.*, 1864) that he has never
observed this species anywhere so large or so finely-
coloured as there. It is found on the hills round
Accrington about the stone walls and old quarries,
though not very numerously, being more plentiful on the
Rossendale uplands. Mr. R. Ramwell says that it is
common near Turton, and Mr. R. J. Howard reports it
as plentiful near Blackburn, though not so much so as
formerly, the enclosing of commons and waste lands
causing the decrease. Writing in the *Magazine of
Natural History* in 1838, Dr. Skaife says that he has
found it most plentiful amongst the mountains in Bowland, and Mr. John Weld informs me that it is still
common on the fells near Leagram Hall, though on the
Wyresdale side Mr. R. Standen considers it rather rare.
Mr. T. Jackson says it breeds plentifully on a large moss
in Winmarleigh, but that at Overton it is not very
common : he finds the birds very fat in spring on their
arrival. It is curious how very early the migrating
species reach the shores of Morecambe Bay ; for example,
in 1882, the Wheatear arrived at Overton on March 23,
whilst at Clitheroe it did not appear till April 12, and it
is so with many others. On Pendle Hill it breeds pretty
numerously, though the nest is difficult to find, and in
my experience, this bird is one of the wariest of its kind.
Occasionally it ventures near the towns, and for many
years a pair had their nest in the wall bounding the
turnpike between Clitheroe and Chatburn, and very near
the former place. On the coast, old rabbit-burrows are
generally selected, the nests frequently being placed
several feet down, and Mr. H. Ecroyd Smith (*Zool.*,

1864) thinks that the eggs there differ considerably from examples procured further inland, being smaller in size, much paler in colour, and also of more delicate texture. They are often seven in number, though generally five or six.

BLACK-THROATED WHEATEAR.

Saxicola stapazina, Vieillot.

The only example of this South-European species which has yet been seen in Britain was shot about the 8th May, 1875, near the reservoir at Bury, by Mr. David Page, and the occurrence was reported in *Science Gossip* for October 1st, 1878, by Mr. R. Davenport, who says: "It is a male bird, in fine mature plumage, and was in very good condition when shot. Its habits, as noticed by several parties for a few days prior to its being captured, were very active, vigilant, and shy. It seemed to hold itself aloof from any of the same order (*S. œnanthe*)."

[This specimen, which is now in the possession of Mr. Doeg, of St. Anne's Street, Manchester, was sent by Mr. Davenport for exhibition at the Zoological Society of London (P.Z.S. 1878, pp. 881, 977); as should always be done in the case of similar rarities. For a concise account of the species, see "An Illustrated Manual of British Birds," p. 23.—Ed.]

GENUS PRATINCOLA.

WHINCHAT.

PRATINCOLA RUBETRA (Linnæus).

LOCAL NAMES—*Eutick, Whin-check, Grass-check.*

A summer migrant, arriving from the beginning to the middle of April, and leaving the latter half of September. In the neighbourhood of Accrington Mr. H. Miller thinks it has got rather scarce, and in Rossendale also, according to Mr. H. Kerr, it nests very sparingly; but everywhere else throughout the county it is common, and in some districts, notably the Fylde, it is abundant. It frequents the cultivated land chiefly, and is very fond of being in a field of newly-cut grass, but Mr. R. Standen says that, in the Goosnargh district, it breeds commonly on the moorlands, as well as in the low country. It lays six eggs, in a nest of dry grass, well concealed in a thick tuft, and is rather a late breeder, the second half of the month of May being the usual time.

STONECHAT.

PRATINCOLA RUBICOLA (Linnæus).

LOCAL NAMES—*Stone-check, Chapper, Flick-tail.*

A resident species, very local, and much less common in winter than summer; the greater part migrating southward about the middle of September or beginning of October, but returning early in the spring. It breeds more plentifully on the sea-shore than elsewhere, and its nest has been taken on the sand-hills at Southport by Mr. T. Jones, at Lytham by Mr. J. J. Hornby, and at Fleetwood by Mr. H. Miller, whilst all round the shores of Morecambe Bay, and especially on Walney, it is fairly numerous. Byerley ("Fauna of Liverpool," 1856) says that it is not uncommon on moorish land, some few remaining over the winter, and Mr. J. F. Brockholes (*Proc. Liverpool Lit. and Phil. Soc.*, 1859–61) writing of birds which then nested within a circuit of ten miles from the Liverpool Exchange, says that it is abundant during the summer months, and affects wastes where gorse grows freely. It also breeds in some of the more elevated districts, but is nowhere common. Mr. J. B. Hodgkinson has seen it in the breeding-season on Beaton Fell, and Mr. John Weld also reports it as nesting on the same range; but near Middleton, where it used to be plentiful, near Accrington, Bacup, and Bury it is now very seldom observed. In the Ribble valley it is met with occasionally, but is rare, and the last example which came under my notice was seen by Mr. T. Altham, near Mytton, in the spring of 1878, being a fine male. It breeds in May, and five is the usual number of eggs.

GENUS RUTICILLA.

REDSTART.

RUTICILLA PHŒNICURUS (Linnæus).

LOCAL NAME—*Red-tail.*

A summer migrant, almost universally distributed, but nowhere very common. It arrives very regularly about the second or third week in April—the males generally being seen a few days before the females come—and leaves the latter half of September. It is seen on migration in all parts of the county, and there are few districts without a pair or two breeding, but Mr. C. E. Reade says that it is very rare at Urmston, and near Clitheroe and Whalley, where it is probably as common as anywhere, its numbers are much fewer than they used to be. It often lays as many as seven or eight eggs, and shows a great preference for the vicinity of farm-buildings, breeding in the orchards and old walls. Its alarm-note is very similar to that of the Chaffinch, but the "chat," with which the Redstart concludes, always serves to distinguish it.

BLACK REDSTART.

RUTICILLA TITYS (Scopoli).

A winter visitor, appearing very rarely. Mr. John Plant reports one shot near Middleton (J. Harrop), and Byerley ("Fauna of Liverpool," 1856) also notes one having been killed near the Dingle in the winter

(Butterworth). Mr. Henry Johnson writes (*Zool.*, 1850, p. 2769) under date January 23rd, 1849, that he examined (presumably at a recent date) a freshly killed specimen which had been shot at Aigburth, near Liverpool, whilst feeding with a small flock of Snow-Buntings.

[One seen near Lytham, F. Brownsword, *Zool.*, 1892, p. 115.—Ed.]

GENUS ERITHACUS.

REDBREAST.

ERITHACUS RUBECULA (Linnæus).

LOCAL NAME—*Robin*.

Resident, abundant, and nesting everywhere. It remains throughout the hardest winters, and, except in severe frost, sings the whole year round. It breeds in the vicinity of the largest towns, but there its manners appear to become demoralized, for in the *Field* of September 10th, 1881, Mr. E. Q. Henriques of Manchester says that a few months before, at Higher Broughton, he found a Robin's nest in a dead cat. Nidification commences early in April, and continues through May, two broods being hatched in the season. The clutches of eggs vary a good deal, and sometimes approach a pure white: their number is usually six, sometimes five or seven.

SUBFAMILY SYLVIINÆ.—GENUS SYLVIA.

WHITETHROAT.

SYLVIA CINEREA, Bechstein.

LOCAL NAMES—*Peggy, Peggy Whitethroat, Cut-throat, Small Straw, Straw-small* (pronounced *Streea-smaw*).

One of the commonest of the summer migrants, breeding numerously. Of all my informants, only Mr. W. A. Durnford reports it—and that in the Furness district—as occurring in but small numbers; in most other places it is abundant. It arrives from the middle to the end of April, and I have occasionally found nests with their full complement of five eggs before May has come in, but this is the usual breeding month. Its habit of perching on hedges and trees by the road-sides, loudly chattering its simple notes, makes the Whitethroat one of the best known of the genus. It leaves again in the latter half of September.

LESSER WHITETHROAT.

SYLVIA CURRUCA (Linnæus).

LOCAL NAMES—*Peggy, Hazel-Linnet.*

A summer visitor; very much less common than the last species, and arriving rather later and departing rather earlier, than it. Mr. A. G. More was certainly misinformed when, in the *Ibis* for 1865, he stated—a statement copied into Mr. H. E. Dresser's "Birds of Europe"—that the Lesser Whitethroat breeds in all counties but Lancashire, &c., &c., for Mr. W. Peterkin, of Clitheroe, avers that he has always found it nesting in that neighbourhood during the last thirty years, and at present it breeds there everywhere in suitable localities. Near Chipping also, Mr. John Weld informs me that it breeds commonly, the nests being in brambles, thickets by the road-side, &c.; one was composed entirely of different dried grasses, without moss or wool, but intermixed with the grass fibres were small quantities of spiders' webs. The inside was exclusively lined with hair, and had six eggs. He also says that the bird seems rather partial to hazel-bushes, hence being called the "Hazel-Linnet." A little further west, a few nests, according to Mr. R. Standen, are seen each year near Haighton and Grimsargh, and this is so, too, about Grange and Leck and in the district south and south-east of Manchester, but elsewhere this species appears to be very rare, though Mr. J. E. Palmer writes me that he observed it near Todmorden in the spring of 1874. At Pleasington, in 1884, Mr. R. J. Howard saw the first nest he had come across in that district. The eggs, five or six in number, are laid the end of May, and the nest is more compactly built than is the case with the other Warblers.

BLACKCAP.

Sylvia atricapilla (Linnæus).

A summer visitor; not appearing till the end of April or beginning of May, and leaving in September. It is very evenly distributed over almost the whole of the county, but is nowhere common, and one or two pairs would probably be as many as could be found in any particular district. Near Liverpool it is rare, and at

Urmston it only occurs on migration, but approaching Manchester on the south side it nests regularly, having formerly been much more plentiful than now. It is of retiring habits, and, away from the towns, is perhaps as numerous as it ever was in recent times, this at least being the case in the neighbourhood of Clitheroe. The eggs, four or five in number, are laid late in May or early in June, and the male will sometimes take a share of the duties of incubation. Mr. Hugh P. Hornby informs me that once at St. Michael's he saw a male unmistakeably *singing* while so engaged. There is very much in common between the Blackcap and the Garden-Warbler, and I do not believe their nests and eggs can be identified of themselves, even by the most practised eye: the notes, too, are so similar that only the best ears can separate them, but the Blackcap's are a little fuller and richer. Both species have the habit of sticking up bits of dried grass in the brambles whose vicinity they are frequenting, as if they had begun building a nest, and then become dissatisfied with the situation; the perfect nest is sure not to be far off. When the young of the Blackcap are fledged, and the old birds bring food to them after they have left the nest, the latter utter sounds exactly like the "mewing" of kittens. It is this bird, together with the Garden- and Sedge-Warblers, which has so often been taken for the Nightingale.

GARDEN-WARBLER.

Sylvia hortensis, Bechstein.

This species was stated by Montagu ("Dictionary of British Birds," 1802) to have been first discovered in

Lancashire, about the end of the last century, and sent from thence to Dr. Latham, by Sir Ashton Lever, of Alkrington Hall, Middleton, the founder of the famous Leverian Museum. Professor Newton, however (Yarrell's "British Birds," 4th ed., p. 415), says the Garden-Warbler "was first made known as a British bird by Willughby, to whom it was sent from Yorkshire by Mr. Jessop of Broom Hall, near Sheffield, under the name of 'Pettichaps.'" It is a summer migrant, and arrives late in April or early in May, leaving again early in September. In the Clitheroe district it is much commoner than the Blackcap, and in the Hodder valley, near Stonyhurst, is comparatively numerous. [Mr. W. F. Brockholes says it is certainly more plentiful than the Blackcap about Garstang.—Ed.] But from all other parts of the county I have it reported as rare, and as only breeding in small numbers, the Liverpool naturalists (*Nat. Scrap Book*, pt. 4) being at issue as to whether it is found at all in their neighbourhood. Dr. Skaife (*Mag. Nat. Hist.*, 1838) considered it very common near Blackburn in 1838, but Mr. R. J. Howard says that now, though generally distributed, it is not numerous. Everywhere else, from Urmston to Barrow-in-Furness and from Blackpool to Rossendale, a pair or two may be found throughout the summer. It lays four or five eggs the end of May or beginning of June, and the nest, as a rule, is shallower and looser than the Blackcap's, being placed usually in brambles, but sometimes in other situations, for Mr. T. Altham has not only found it in nettles, and in a fern, but in one instance on the branch of a plane-tree, 14 feet from the ground.

GENUS REGULUS.

GOLDCREST.

REGULUS CRISTATUS, Koch.

Resident, and having its numbers very much increased on the approach of winter by immigrants from the north. Its remaining in summer is conditioned by the presence or otherwise of woods of the spruce-fir, to whose branches it attaches its pendulous nest, for, although Mr. John Weld says he has seen the nest on the yew-tree, and Mr. J. B. Hodgkinson that he has found it among whins, the spruce is almost invariably selected. The severe winters and springs of 1879 and 1880 almost destroyed it about Chipping, and all down the Hodder its numbers were sensibly decreased, but in 1882 the normal quantity was recovered and Longridge Fell tenanted as numerously as ever. In the south of the county it is rare in summer, and near Bury it appears chiefly in winter, but the report of the Natural History Society of that town for 1871 states that the nest and eggs have been taken at Cockey Moor and Reddish woods. In the wooded districts bordering the Fylde it breeds sparingly, as also near Accrington; and on Billinge Hill, within $1\frac{1}{4}$ miles of the centre of the populous town of Blackburn, its nest may occasionally be found. Mr. John Hardy has never seen the nest in the county, but has observed *pairs* in the wooded parts of Heaton Park during the breeding-season, and the species is also seen in Cliviger in the woods at Holmes Chapel. Mr. T. Jackson says that he has shot a few at Overton for several years back, always in the autumn, and that he thinks it is increasing in numbers. In Furness, Mr. W. A. Durnford reports it as resident, but not common. A full nest contains eight or nine eggs, and these are usually laid by the 1st of May. The nest is invariably very near the end of a branch, and is placed at various heights; sometimes in the lowest, and sometimes in the very highest branches.

FIRE-CREST.

Regulus ignicapillus (C. L. Brehm).

The only instance of the occurrence of this rare winter visitor is supplied me by Mr. John Hardy, of Manchester. He says, " once seen and three specimens obtained in Hough End Clough and in a small wood with a few Scotch firs near to it, in the month of December, 1851; its habit and call-note seemed exactly the same as that of the common species, with which it became occasionally mixed. I could not ascertain that the flock had been seen in any other place either before they came, or after their departure. I had heard of them having been seen in the same neighbourhood the year before, but I could not get sight of a specimen which had been collected: the birds seen at the date above given were undoubtedly Fire-crests." [It does not appear that the above specimens have been identified by any authority whose decision would carry weight. See also Mr. J. H. Gurney's remarks in *Zoologist* for 1889, p. 174.—Ed.]

GENUS PHYLLOSCOPUS.

CHIFFCHAFF.

PHYLLOSCOPUS RUFUS (Bechstein).

LOCAL NAME—*Peggy.*

A summer visitor, rarely seen except in one or two localities, and nowhere common. It is not mentioned by Mr. John Blackwall in the tables of migrants, having reference to the north side of Manchester, published in his "Researches in Zoology," but on the south side Mr. John Hardy says that it was formerly not a rare bird, arriving very early, and breeding regularly, now being certainly less common near the city, apart from the fact that its haunts in this special district have been made less retired, and in some cases destroyed. In the Clitheroe district, Mr. W. Peterkin never heard it, and before 1877, when Mr. T. Altham heard one in the Hodder valley, a long interval had passed without its appearing. The last-named observer heard it also near Clitheroe in 1879 on April 2nd, and in 1880 on June 2nd, and took a nest of five eggs near Habergham Eaves, Burnley, in 1871, where, too, he had occasionally heard it in other years. I myself have never come across it during fifteen years of pretty close observation. At St. Michael's-on-Wyre it is an irregular visitor, Mr. Hugh P. Hornby having seen it in the springs of 1880 and prior years, but not since. Mr. J. B. Hodgkinson says that it breeds every spring near Preston, and Mr. R. Standen also finds its nest near Goosnargh, where it arrives about March 20th, and leaves about October 1st. Mr. John Weld says that it arrives the end of March at

Leagram Hall, and is said to come most years. At Grange it breeds regularly, according to Mr. J. B. Hodgkinson, who has often shot it there in spring, and on the Cumberland border, Dr. C. A. Parker of Gosforth says it is found breeding. In its habits it resembles very much the Willow-Warbler, but its nest is much more difficult to find. The two species are often confounded, though the call-note of the Chiffchaff is very distinct, and the late Mr. Thomas Garnett of Clitheroe, a gentleman who had an exceedingly good ear, used to say that he could distinguish the difference between the first and second notes, the "chiff" and the "chaff."

WILLOW-WARBLER.

PHYLLOSCOPUS TROCHILUS (Linnæus).

LOCAL NAMES—*Peggy, Peggy Whitethroat, White Wren, White Robin, Sweet Willie, Tomtit, Milly Thumb, Willow-Wren.*

The commonest of the summer migrants, and abundant everywhere; usually arriving the first week or two of April, but sometimes being seen the last week of March. The majority leave in September, but laggards often remain till October. It is remarkable with what unanimity this species begin its song after getting to its breeding-haunts. One day the woods shall be comparatively silent, and the next every hedge and every clump of trees shall be full of sweet melody, the residents vying with the newcomers as to who shall be loudest and longest in their strains. It is a disputed point whether summer migrants begin to sing immediately on arrival, but I think it probable they do if the weather be fine; without hearing the song of a species like the present, it would be easy to miss its coming for some time. Although the nest is usually placed on the ground, it is found pretty often in a low bush where there is not much thick grass, and Baron von Hügel reports (*Zool.*, 1872) an instance where he saw a nest placed on the extremity of a branch of a small fir, 16 feet from the ground: Mr. T. Altham also once found one wedged on the top of two branches of a spruce fir, 14 feet from the ground. The nest is domed at first, but when the young are hatched, their weight soon makes the dome disappear. The bird shows great attachment to its home, and in Rennie's *Field Naturalist* for April 27th, 1833, a correspondent signing "Rose, Blackburn," relates how one insisted upon returning to the nest, repairing it, and laying more eggs, after a lot of ducks had pulled it in pieces. Mr. Thomas Fry of Liverpool (*Nat. Scrap Book*, pt. 3) gives a similar incident, where a Willow-Warbler stuck to its nest after it had been demolished by a terrier. The eggs are seven or eight in number, and are usually laid in May; the spots on them are red, in contradistinction to the purple-black ones on the Chiffchaff's. The two species may almost always be separated (Seebohm, *Ibis*, 1877, p. 66) by the varying lengths of the primary quill: in the Willow-Warbler the second is intermediate in length between the fifth and sixth, but in the Chiffchaff it is considerably shorter than the sixth.

WOOD-WARBLER.

Phylloscopus sibilatrix (Bechstein).

Local Names—*Wood-Wren, Fell-Peggy.*

A summer migrant: not so common as the Willow-Warbler, but much more plentiful than the Chiffchaff. Its times of arrival and departure are probably about as stated by Blackwall ("Researches in Zoology," p. 6), viz., April 28th and September 6th, varying little from these. Its favourite localities are those in which plenty of wooded heights occur, and in the south of the county it is seldom seen. Byerley ("Fauna of Liverpool," 1856), says that it is rare near Liverpool, and south-east of Manchester Mr. John Hardy has not seen a nest since 1849, when there were two pairs breeding in Hough End Clough. It is found at Prestwich and Middleton, and breeds rarely in Rossendale. It is common and breeds between Preston and Pleasington; and about Haighton, though rare, it occurs regularly. In Furness, Mr. W. A. Durnford says it breeds numerously. At Read and Huntroyde, near Padiham, it is common, though not so much so as formerly, and its nest is occasionally found on Whalley Nab. Near Clitheroe it breeds every year in the woods at the foot of Pendle Hill, but Longridge Fell and the wooded banks of the Hodder are its favourite habitat, and there it may almost be called numerous. The nest, loosely made of dry grass, is always well concealed among either dead leaves or growing herbage, and of itself, would be very difficult to find; but if the piece of ground near which a cock bird is singing be rapidly tramped over so as to disturb the hen from the nest, she immediately begins to utter her " sorrowing " note, and a careful watch of a quarter of an hour or so will generally suffice to see her back to it. The number of eggs is six, sometimes seven, and they are laid during the last three weeks of May.

GENUS ACROCEPHALUS.

REED-WARBLER.

ACROCEPHALUS STREPERUS (Vieillot).

A summer visitor, but rare and local; swamps and reed-beds, such as it frequents during the breeding-season, being few and far between in Lancashire. Mr. J. F. Brockholes (*Nat. Scrap Book*, pt. 8) says he has known the nest in a reedy ditch between the Maghull railway station and Sefton meadows, and, from hearsay, that the bird sometimes occurs in the osier beds around Warrington. In his " Harmonia Ruralis," 1794, Bolton writes that he has had birds sent him shot on the river Roch, but it is rarely seen now in the Bury district. Mr. W. Peterkin has known of its nesting near Manchester, and Mr. David Mitchell near Morecambe, and Mr. R. Standen thinks a pair or two annually visit the Goosnargh district, he having in June, 1878, found two nests there. Seven or eight years ago, Mr. Louis H. Simpson tells me a nest was taken near Preston in June by Mr. Richard Sharples, and I have seen the eggs in the former gentleman's collection. The nest was in a privet hedge which divided Mr. Sharples' garden from a small stream, and was a good height up. Elsewhere, I have no notices of its presence, nor am I furnished with any dates of migration, but the probability is that it comes and goes with the other members of the family of Warblers. I have had large experience of its habits in other places, and have usually found it breeding in the outside portions of reed-beds, though sometimes in a willow-fork, the nests being almost entirely of fine grass, the flowering tops much used, the lining the finest, sometimes intermixed with a little moss and wool; they are usually one or two feet above the water level. The Cuckoo very often lays its eggs in the nests of the Reed-Warbler, though they are so deep and narrow that I think its bill must be used for the purpose of deposit. The normal number of eggs is four.

SEDGE-WARBLER.

ACROCEPHALUS PHRAGMITIS (Bechstein).

A summer migrant; arriving the last week in April or the first in May, and leaving in September. It is one of the best known of the Warblers, and from its habit of singing almost the night through, and from its imitative powers, in many places gets called the Nightingale, and in others the mock Nightingale, or the mocking-bird. It is indeed a wonderful mimic, but its habit of mixing up its own song with that of the birds it is imitating always discovers the true performer. Mr. Thomas Garnett has some notes on this subject in the *Magazine of Natural History* for 1832 and 1834, which are worth reproducing. He says, "I have heard it imitate in succession (intermixed with its own note of *chur chur*) the Swallow, the House-Martin, the Greenfinch, the Chaffinch, the Lesser Redpole, the House-Sparrow, the Redstart, the Willow-Wren, the Whinchat, the Pied Wagtail, and the Spring Wagtail; yet its imitations are confined to the notes of alarm (the fretting notes as they are called here) of these birds, and so exactly does it imitate them, both in tone and modulation, that if it were to confine itself to one (no matter which) and not interlard the wailings of the little Redpole and the shrieks of the Martin, with the curses of the House-Sparrow, the *twink twink* of the Chaffinch, and its own *care for nought* chatter, the most practised ear would not detect the difference." He also says that he has heard it mimic, and not invariably the alarm-notes, the Starling, Whitethroat, and common Linnet, but never Larks or Thrushes, the notes of the Sparrow, Whinchat, Swallow, and Starling being its chief favourites for practise on. Bolton ("Harmonia Ruralis") states in 1794 that in some parts of Lancashire it is taken for the Nightingale, and that it is plentiful, inhabiting the borders of still ponds, and marl-pits, and this is still true, for it is found on the lower levels wherever there is a suitable marshy spot of ground, and breeds numerously in such localities throughout the whole county. Up to the year 1861, Mr. John Plant says that it bred in the osiers on the banks of the river at Peel Park in Salford. Its nest is sometimes partly suspended after the manner of the Reed-Warbler, but more often is supported in the usual way, and occasionally it is built in a hedge, if pretty thick, seven or eight feet from the ground. The eggs, six or seven in number, are laid from the middle of May to the middle of June: Dr. St. Clair tells me that on May 9th, 1879, he saw two nests near Blackpool each with five eggs, and Mr. R. Standen has twice found the nest with a brood of young in September, but these dates are both extraordinary. The white streak over the eye is the most distinguishing part of the plumage.

GENUS LOCUSTELLA.

GRASSHOPPER-WARBLER.

LOCUSTELLA NÆVIA (Boddaert).

A summer visitor, whose whirring note is usually heard the last few days of April or beginning of May. Its retiring habits make it a difficult subject of observation, but it probably leaves with the rest of the Warblers in September. In suitable localities it is not uncommon, but is nowhere abundant. In the Manchester district it nests regularly, according to Mr. John Hardy; [also Mr. Frank Nicholson, Ed.]; and near Garstang, Mr. J. A. Jackson generally hears one or two each summer. It breeds in the Goosnargh neighbourhood, where it is called "Hurrer" or "Huzzer"; and Mr. John Weld says it was very common near Chipping in 1882. It is probably more numerous in the Ribble valley than anywhere else, and near Clitheroe many pairs breed each year; it being found also annually at Balderstone, and within three miles of Preston Mr. J. B. Hodgkinson considers it common. On Longridge Fell, above Dutton, it nests at a considerable altitude. Mr. Thomas Jackson records its appearance at Overton in June in 1879 and succeeding years, and Mr. W. A. Durnford's informant W. B. K. ("Birds of Walney," 1883) states that it occurs on the mainland of Furness. Its favourite habitat is the thick undergrowth of a young plantation, and it will frequent such a place for years until the growing timber has rendered the means of concealment too scanty, when it is entirely deserted; an open meadow is also often chosen for a breeding-place. The nest is very well concealed, but there has been great exaggeration as to the difficulty of finding it, and I only know of one instance when even long and continued searching failed to discover its position. In my experience, the tortuous passage (we read of) through thick grass is a myth, and the bird also leaves its nest quite as often by a low flight as by slipping over the edge and creeping away in mouse fashion. The nest is composed entirely of rough grass, without any difference of texture between the out- and the in-side, and is generally hidden deep amongst the tall grass, or in a tuft in the middle of a low blackberry-bush. A pair have for several years nested within twenty yards of my garden, and on one occasion I was enabled to crawl within two feet of where a bird was trilling. It was singing about the middle of a four feet hedge, worming about among the branches, and occasionally changing its position slightly. This movement, and the continuous character of the note, with an occasional echo, are no doubt the causes of an apparent ventriloquial power, with which it has sometimes been credited: quite wrongfully, in my opinion, for where there is no echo, the position of the bird may always be accurately found, and at once. The note varies very much in depth, and is uttered in fine weather usually from about 8 P.M. to 2 A.M., almost continuously and with only the slightest break. This species rarely sings in the day-time, and either the male assists in incubation, or the female also has the power of trilling, as I have heard a bird utter the note immediately after leaving the nest. During execution, the beak is held wide open, and the mandibles motionless, but the tongue appears to quiver. A full nest contains six eggs, and these are laid during the month of May, oftenest about the 15th, though I have seen young, a week old, on the 4th of June. Incubation sometimes begins before the full complement of eggs is laid, as I have found to happen in several instances.

SUBFAMILY ACCENTORINÆ.—GENUS ACCENTOR.

HEDGE-SPARROW.

ACCENTOR MODULARIS (Linnæus).

LOCAL NAMES—*Dykie, Dykie-Sparrow, Dunnock, Dunny, Hedge-dunny.*

A resident species, and common, where there is any cover, throughout the whole county. It remains the hardest winters, not often more than two or three being seen together, and is very tame and sociable, keeping with the fowls about the barns. In favourable weather it will sing at any period of the year. It is double-brooded, and the well-known blue eggs of the first brood are laid from the middle of April to the first week in May, and are four or five in number.

FAMILY CINCLIDÆ.—GENUS CINCLUS.

DIPPER.

CINCLUS AQUATICUS, Bechstein.

LOCAL NAMES—*Water-Ouzel, Water-Crow, Bessie-dowker, Betty-dowker.*

Resident, and remaining on the inland streams even in hard winters, though in the very severest—as, for instance, that of 1878-79—it disappears. It is essentially a bird of the hill-district, and is hardly known in the flat south-west portion of the county. It is frequent, according to Mr. John Hardy, in places on the large moors about Bury and Oldham, and northward, becomes increasingly numerous as the manufacturing districts are left behind. Corry ("History of Lancashire," 1825) says that it nests on the rocky shelves overhanging the Spodden, near Rochdale, and it still occasionally breeds in Rossendale. It is mentioned (ed. Raines, Chetham Society, 1848) by Nicholas Assheton of Downham in his diary, who relates how on November 4th (1617) he went "downe to the water" and "Sherborne killed a water-ousle," and on the rocky streams of Ribblesdale it is abundant at the present day, as also of all the fell country northwards to Furness and the borders of Windermere. The nest is placed under the overhanging banks of, or in crevices of rocks near, brooks and rivers, on ledges behind waterfalls, or on jutting stones under bridges, and is invariably lined with leaves. Particular spots are frequented year after year even if the nests are robbed. An interesting account of the construction of the nest is given by Col. H. W. Feilden in the *Zoologist* for 1867. Writing from Feniscowles he says, "Under the arch of a bridge, over which the high-road passes, a pair of Water-Ouzels are now nesting. They commenced building about March 1st; the very cold weather that commenced about the 6th stopped further proceedings, but building was resumed on the 24th, the very day a thaw commenced: by the 31st the outside of the nest was completed, and on the 7th of April there was one egg in. I noticed very particularly each day the progress made on this nest: the birds commenced building from the bottom, and then piled a ring of moss, in the shape of the letter

O, against the wall; they then laid moss alternately on one side and the other, and by the time the ring of moss was completed, the base of the nest protruded four or five inches, and the top about one inch from the wall, the thickness of the walls of the nest also tapering off from bottom to top. When the ring was completed, as I have described, the Ouzels changed their tactics, and commenced building down from the top until the whole of the nest was arched over, the entrance being placed over the stream more at the base of the nest than the side. It is wonderful how so large and heavy a structure as this clings to the wall, for where the nest is placed there is only a slight convexity in the face of the stone, hardly appreciable to the eye when the nest is away." I think, however, that there must have been some slight projection, perhaps covered before noticed by the observer, on which the base of the nest might rest. Mr. Thomas Garnett has also some pertinent remarks on this species in the *Magazine of Natural History* for 1834. He says: " The Water-Ouzel does sing very frequently, and as much in the winter as at any time. Perched on a stone, or a piece of ice, it chirps away at a famous rate; but its song consists almost entirely of its note *zeet-zeet*, which it hashes up in all sorts of ways." He notes its resemblance to the Wren in its habits and motions, its nods and curtsies, and the cocking of its tail, and respecting its power of walking under water, continues: " I have repeatedly seen it doing so from a situation where I had an excellent opportunity of observing it, the window of a building directly over the place where it was feeding. It walked in, began to turn over the pebbles with its bill, rooting almost like a pig, and it seemed to have no difficulty whatever in keeping at the bottom, at all depths where I could see it: and I have frequently observed it when the water just covered it, and its head appeared above every time it lifted it up, which it did incessantly; turning over a pebble or two, then lifting its head, and again putting it below to seize the creepers (larvæ of insects) it had disturbed. Besides, its speed was too slow for diving. Every aquatic bird I know moves much faster when diving than when either swimming or walking, and its course is generally in a straight line, or nearly so: but the Water-Ouzel, when feeding, turns to the right or left, or back again to where it started, stops and goes on just as it does when out of the water. Yet, when it wished, it seemed to have the power of altering its own gravity, as, after wading about two, or perhaps five, minutes, where it could just get its head out, it would suddenly rise to the surface and begin to swim, which it does quite as well as the Water-hen. The awkward, tumbling, shuffling wriggle is occasioned by the incessant motion of its head as it turns over the gravel in search of creepers, which, it appears to me, form the whole of its food." He believes it is catching creepers when supposed to be devouring salmon-spawn and goes on: "If this were the case (and it is a fact well worth ascertaining) it was rendering an essential service to the fisheries . . . because these creepers (the larvæ of the May-fly, bank-fly, and all the *drakes*) are exceedingly destructive to spawning beds, and as the Water-Ouzel feeds on them at all other times, and as they are more abundant in the winter than at any other season, I think this is the more probable supposition." The Dipper, however, certainly feeds on fish sometimes, for on July 11th, 1879, I disturbed a bird from a nest beneath which was quite a heap of young minnows, and Mr. T. Altham has observed a similar instance, they probably having been refused by the young. It is an early breeder, and sometimes lays its eggs in the beginning of March. The earliest date I know of for eggs is the 1st of March, on which day in 1880 a nest of five was seen by Mr. R. Standen, and within a week or two of this I have known and heard of several instances. The bulk, however, are laid the beginning of April, and the number is four or five. Mr. J. P. Thomasson (*Zool.*, 1861) thinks four are oftener laid than five, and says that he once found a nest containing only three young. A very curious circumstance in connection with this species was communicated to the *Field* of May 13th, 1871, by Mr. Louis H. Simpson of Preston. He writes, "On Saturday last I found on one of our rivers three Water-Ouzels' nests, one above the other, the roof of the lowest one being the bottom of the next, and so on. The two lower ones had three eggs apiece, which were quite fresh; but the top one contained four young birds just hatched." Mr. Simpson has since informed me that the nests were under an overhanging bank on the river Brock, near Garstang, and that the eggs are now in his collection. The Dipper is double-brooded, and brings up two sets of young in the season.

FAMILY PARIDÆ.—GENUS ACREDULA.

BRITISH LONG-TAILED TITMOUSE.

ACREDULA ROSEA (Blyth).

LOCAL NAME—*Bottle-Tit.*

Resident, and very generally distributed, but much commoner in some districts than others, and, on the approach of winter, having its numbers considerably increased everywhere by migrants. Byerley says that it breeds near Liverpool, and is not uncommon in the winter time, flying generally in families of from eight to twenty. On the eastern side of the county, where timber is scarce, it is rarely seen except on migration, but in many wooded districts it breeds in some numbers, being most common perhaps in the valleys drained by the Ribble and Wyre. Its beautiful nest is generally fixed in high hedges, often very close to footpaths, and not seldom on the lower branches of trees, placed in some recess, and covered all over with lichen. In several instances I have seen the nest, on branches of trees overhanging the river Hodder and growing out of its steep banks, 50 feet above the level of the stream. This Titmouse is an early breeder, and lays its ten or eleven eggs the first two weeks of April.

GENUS PARUS.

GREAT TITMOUSE.

PARUS MAJOR, Linnæus.

LOCAL NAMES—*Ox-eye, Ox-eye Tit, Tom-tit, Nope, Tommy-nope, Billy Biter, Black-cap.*

Resident and plentiful both in summer and winter. This and the three following species have very much in common in their habits; all of them breeding in holes of trees, and all boring a hole for themselves, a perfect circle, if a suitable natural one be not ready to hand. Their call-note too is the same. The Great Tit also breeds in old walls, and when it leaves the nest usually covers the eggs with pieces of wool. When on the nest, this species and the Blue Tit are exceedingly reluctant to leave the eggs, it being often necessary to push them off by main force, when they retire to another part of the cavity, hissing loudly, and sometimes making furious dashes at the spoon and lighted vesta used for convenience in extracting the nest contents. In 1878 Mr. T. Altham found a Great Tit sitting on its nest with no eggs under it. It was close to where he was working, so he visited it every day for three weeks, the bird always being on. At each visit he felt under her for eggs, but never found one, and supposing that a mouse took every egg as it was laid, he set traps, but caught nothing. At the three weeks' end he had to go away, so caught the bird, and on dissection it proved to be a female, whose fully developed ovary is still in my possession. In this year too a Blue Tit was doing the same thing, but she only sat on her empty nest for one week, and then deserted it. The Great Tit lays seven to nine eggs usually, though I have seen as many as twelve, and the nest is completed about the 1st of May.

BRITISH COAL-TITMOUSE.

Parus britannicus, Sharpe and Dresser.

Resident, and commonly seen almost everywhere in autumn and winter, in small flocks with the other species of the genus; but much more local when breeding, and less numerous altogether than the Great and Blue Tits. Mr. John Hardy has only once seen it breeding near Manchester, namely, at Barton-on-Irwell, and Byerley records it as only an occasional winter

visitor to the Liverpool district, but at Urmston Mr. C.
E. Reade considers it more numerous than the Great
Tit, it nesting chiefly in holes of trees, but also in walls.
Northwards, a pair or two are found in most of the
suitable woods, and Mr. R. J. Howard has noted it as
more plentiful near Hawkshead than elsewhere. It
nests regularly in the woods on Longridge Fell, where
it almost invariably chooses a hole in a tree within two
feet of the ground. oftenest under the roots, building
with moss, and lining chiefly with down mixed with a
few hairs. It commences to lay its eggs the last days
of April, and they are generally ten or eleven in number.

MARSH-TITMOUSE.

PARUS PALUSTRIS, Linnæus.

Resident, and, like the Coal-Tit, more generally distributed during winter, though only in small numbers
anywhere. Except in the Furness district, from which
I have no notice, it appears to be found in summer in
most of the suitable localities, and in places it is stated
to be fairly plentiful: as for instance Chipping, Preston,
and Balderstone. In the neighbourhood of Clitheroe it
had not been noticed for some time until 1876, when
Mr. T. Altham found a nest by the Hodder; but some
years before it had been identified at its nest in the
same valley by Mr. W. Peterkin and Mr. J. P. Thomasson, and the late Mr. Thomas Garnett in his time too
used to find it. The last-named gentleman wrote in the
Magazine of Natural History for 1832 detailing the
habits of the bird when breeding, and spoke of the
tenacity with which it holds by its nest, its hissing, and

pecking at a stick or anything else which may be inserted in the hole, which have been noticed as characteristic of the Great and Blue Tits. It breeds in holes in trees, generally near the ground, in old stumps and gate-posts, the thatch of huts and hovels in waste places, and sometimes in walls, and lays about nine eggs the first week in May.

BLUE TITMOUSE.

Parus cæruleus, Linnæus.

Local Names—*Nope, Blue Nope, Mope, Blue Mope, Tom-tit, Tit-nope, Tom-tit Nope, Jitty-ja.*

Resident, and everywhere the commonest of the Tits at all seasons of the year. It nests in all sorts of curious situations, appropriating any old stump with a suitable hole in it, and readily taking to boxes if specially fixed for its use. A Manchester correspondent of the *Mag. of Nat. Hist.* for 1832 closely watched some Blue Tits breeding in boxes he placed for them, and before incubation began, the eggs were always covered in the morning, after the fresh one was laid. The fæces of the young were always carefully removed as soon as voided. In one instance the female was killed after the young were hatched, and they were forsaken by the male and died. Very different this from a Long-tailed Tit observed by Mr. Thomas Garnett, which (its mate being destroyed) brought up a numerous family by its own unwearied exertions. The Blue Tit begins to lay its eggs about the 1st of May, and they are usually ten in number, though I once took thirteen from a single nest. It is very fond of feeding on the insects which frequent the alder.

FAMILY SITTIDÆ.—GENUS SITTA.

NUTHATCH.

SITTA CÆSIA, Wolf.

The Nuthatch is an exceedingly rare species in Lancashire, and is very seldom seen. It has, however, been known to breed, and has also been observed in autumn and winter. In Mr. A. G. More's paper in the *Ibis* of 1865 on "The Distribution of Birds in Great Britain during the Nesting Season," it is stated that Mr. C. S. Gregson has seen it in the woods of Wyresdale, and its occurrence here formerly is also noted in the MSS. of the late Rev. J. D. Banister. Mr. John Hardy writes me that he has once known it breeding near Manchester, several times having seen two or three in a day in the nesting season, and having also observed it oftenest early in spring, though by no means regularly at any season. The last occurrence I have come across is of one which killed itself in September, 1880, against the glass of the vinery at Waddow, near Clitheroe, the residence of Mr. James Garnett, who has had the bird preserved: it had been seen about for two or three weeks previously.

FAMILY TROGLODYTIDÆ.—GENUS TROGLODYTES.

WREN.

TROGLODYTES PARVULUS, Koch.

LOCAL NAMES—*Chitty, Chitty Wren, Kitty Wren, Jenny Wren, Tom-tit.*

Resident, and everywhere abundant, remaining through the hardest winters, though those of 1878–79 and following years sadly thinned its numbers. At this season the birds may be heard calling to each other in the evenings, and then all go to some old nest, Martin's or other, and huddle together for warmth whilst sleeping. The Wren frequents the neighbourhood of the largest cities, and Byerley states that it is common near Liverpool, whilst throughout the Manchester district it is still plentiful. It is very pugnacious, and in the *Zoologist* for 1869, under date March 17th, Mr. J. Murton, of Silverdale, gives an account of an extraordinary contest he once witnessed. He says: "I was walking past an ivy-clad rock, when my attention was attracted by a rustling among the leaves, and in a few moments down came the objects which were causing the disturbance, in the shape of two Wrens closed in desperate conflict. They continued the combat at our feet, and we managed to capture one of them under a hat, the other making its escape to the top of the rock, and immediately giving out its vigorous notes of defiance. On our prisoner being released, he forthwith returned the challenge in notes equally loud, and in less than a minute the two had again closed, and again came to the ground struggling together. A second time one of them was caught under the hat, but it got away, and lost no time in answering the note of battle already sounded by its antagonist. The contest was resumed for the third time, and with the same result, the two falling to the ground together as before. I attempted another capture, but failed. Whether the defiant notes which were again uttered ended in a fourth battle, I did not stop to ascertain. I noticed that in closing they grasped each other's feet, and fought with their beaks." The well-known habit of the Wren to build several other nests, generally called *cock-nests*, in the vicinity of the one it chooses for incubation, and to leave them unlined, still remains without satisfactory explanation. Writing in the *Mag. of Nat. Hist.* for 1830 and 1832, Mr. Thomas Garnett says that, on the 2nd of May in the latter year, he knew of a dozen cock-nests, which had remained in the same state since the middle of April, other nests, found about the same time, having young. If the first nest be taken, the birds will occasionally take possession of a cock-nest, as he has found such an one, after remaining unfinished for several weeks, fitted with a lining and used. He once pulled out a nest already lined, and the birds immediately occupied an adjacent cock-nest, and a lot of young, which had left their original home, were found roosting in a similar one hard by. I have usually found the nest which is preferred more carefully concealed than the cock-nests, but not invariably so, and the materials are often quite out of unison with the surroundings. The song of the Wren is the most powerful I know, relatively to the size of the bird, and is not rarely heard in winter. The six, or sometimes seven, eggs are laid the first days of May, and the nest is always lined with feathers if these are plentiful, if not, with cow's hair mixed.

FAMILY MOTACILLIDÆ.—GENUS MOTACILLA.

WHITE WAGTAIL.

MOTACILLA ALBA, Linnæus.

The White Wagtail probably occurs much more frequently in Lancashire, as well as other places, than is supposed, its great similarity to the Pied Wagtail making it very liable to be overlooked by the ordinary observer.

But, in a very large proportion of the collection of birds made by the cottagers, and used to ornament their dwellings, examples of this species, both in summer and winter dress, may be seen, and a more thorough appreciation of the differences in plumage between it and the commoner bird, would no doubt increase very much the number of recorded occurrences. In Hewitson's "Eggs of British Birds" (3rd edit., 1856), Mr. Samuel Carter of Manchester says that he has seen White Wagtails near that city, generally in ploughed fields, and rarely by the side of water, and continues: "I have also seen them at Turton, a small village between Bolton and Blackburn, and though there is a large lake of water in the neighbourhood, in which I have frequently fished, I never saw one of these birds by its edge, but frequently the pied." The same gentleman, writing to the *Zoologist* (1857, p. 5517), also says that it appears near Manchester in spring and autumn, but that he has never heard of its breeding in the neighbourhood, nor has he met with it in winter. Mr. John Hardy, who, with a collector named Edward Jacques, accompanied Mr. Carter when the nest spoken of by Hewitson was taken at Holme (not Whittlesea), in Hunts, tells me that it occurs every summer in the district south-west of Manchester, and that several good specimens are in the collection at Queen's Park there. He says that Jacques, who was a clever mounter of small birds, distributed many White Wagtails shot by himself and others. I am informed by Mr. R. Davenport that one was killed at Bradshaw Fold, near Middleton, by Mr. J. Holland, in April, 1870, and another on July 8th in the same year near Bury reservoir by Mr. F. Oates. Mr. J. B. Hodgkinson has no doubt of its breeding near Preston, and says there is a specimen in the museum there shot on the Ribble below the town by Mr. James Cooper. He saw a pair in 1880, in the breeding-season, above Dutton, higher up the Ribble valley. [Mr. R. J. Howard says that he meets with a few every spring on Tarleton Moss.—Ed.] I have carefully observed the White Wagtail in Norway and Holland, but can find no difference from the Pied Wagtail either in habits, mode of nidification, colour of egg, or song; but the plumages are very distinct, the back of the latter being in summer black, whilst that of the former is ashy-grey, and the Pied Wagtail is also considerably darker at all other seasons of the year.

PIED WAGTAIL.

MOTACILLA LUGUBRIS, Temminck.

LOCAL NAMES—*Water-Wagtail, Willie Wagtail.*

Resident and common, breeding everywhere; its numbers in winter being greatly reduced by migration, though there are few districts where a pair or two may not be seen the year round. It collects in considerable numbers, with the Yellow Wagtail, in spring and autumn, in bushes overhanging ponds, and in the month of September many hundreds are sometimes seen flocking together in the evening, probably for migration purposes, as the majority appear to be birds of the year. It is not uncommon on the sand-hills of the coast, and here Mr. H. Durnford (*Zool.*, July, 1873) says it generally nests close to one of the numerous pools under the shelter of some overhanging tuft of grass. It is double-brooded, and lays the first clutch of from four to six eggs the end of April or beginning of May.

GREY WAGTAIL.

MOTACILLA MELANOPE, Pallas.

LOCAL NAME—*Rock-Wagtail.*

Resident; breeding on most of the secluded and rocky streams in the northern parts of the county, and in winter, with the Pied Wagtail, often appearing close to the towns and villages, especially where there are streams in which sewage flows; these being both rich in insect life and having the advantage of remaining open in the severest weather. Near Liverpool, and in the other low-lying districts, it is seldom seen except on migration, and, as a breeding species, is not anywhere common, being more numerous on the Hodder than any other place I know. Here each pair seems to appropriate a certain stretch of river, and it is very unusual for two nests to be seen in any close proximity. The nests are built of moss chiefly, and lined with hair, and are placed on ledges of rock, seldom more than a few yards above the level of the stream, and often partially concealed by trailing branches of ground-ivy. The female sits very closely, and makes a great to-do when disturbed, uttering her alarm-notes, and flying from stone to stone, and from tree to tree, with the greatest restlessness and anxiety. She lays her first set of eggs (four or five in number, rarely six) by about the second week in April, being one of the earliest breeders, and later on another brood is hatched.

YELLOW WAGTAIL.

MOTACILLA RAII (Bonaparte).

LOCAL NAMES—*Yellow Water-Wagtail, Yellow Land-stir* (sometimes pronounced *Lawnster*), *Seed-fool, Seed-fore.*

A summer visitor, arriving about the second week in April and leaving in September. From almost all parts of Lancashire it is reported to me as common, even in the neighbourhood of the largest towns, and the cultivated land is its favourite habitat, it breeding oftenest in corn, hay, and fallow fields. Mr. Hugh P. Hornby, however, thinks its numbers are less than they were six or eight years ago at St. Michael's-on-Wyre, and Mr. T. Jackson also considers it rather rare between the Lune and Morecambe Bay. It derives its local names from appearing at sowing-time in spring, and from its habit of following the plough, and feeding on the insects turned up by the share. Building oftenest among the growing crops, its eggs are not found in anything like proportion to the abundance of the birds, and apart from this, the nest is generally well concealed among the rough clods. It lays five or six eggs, and is rather irregular in its times of nidification, sometimes beginning to sit as early as the 26th of April, but usually quite a month later.

GENUS ANTHUS.

MEADOW-PIPIT.

ANTHUS PRATENSIS (Linnæus).

LOCAL NAMES—*Tit-lark, Ground-lark, Cheeping-lark, Swinpipe.*

Resident, and probably the commonest bird we have; breeding numerously everywhere, from the sand-hills just above high-water mark to the tops of the loftiest hills, and equally so on the most cultivated and the most waste lands. In severe winters it is partially migratory in the bleaker districts; for instance, Rossendale, where, as Mr. Kerr says, owing to the comparative absence of trees and cover, many of the so-called resident birds are migratory. Mr. T. Jackson says that on the marshes adjacent to the Lune estuary this bird is a constant visitor in great numbers during open weather in winter. Its nest is chosen far more frequently than any other by the Cuckoo, and seldom contains fewer than six eggs, which are laid through May and June.

TREE-PIPIT.

ANTHUS TRIVIALIS (Linnæus).

LOCAL NAMES—*Ground-lark, Tit-lark.*

A summer visitor, whose quiet plumage, and want of other very noticeable characters make its arrival not so much remarked as that of many migrants. Blackwall gives April 14th as the average date of its arrival at Crumpsall for the fifteen years from 1814 to 1828, but most observers note it a fortnight later than this: September 13th is his date for its departure. It is fairly plentiful in all wooded districts, and is very generally distributed over the whole of the county, being proportionately rarer where timber is scarce. Mr. W. A. Durnford says that in Furness it occurs in flocks, chiefly at the migratory season. Woods, and banks in their vicinity, are its favourite breeding-places; the five or six eggs (one variety among many of which very much resembles those of the Wood-Lark) are laid about the middle of May. Its short and little varied song is chiefly uttered when, from the branch of a tree, it rises a yard or two into the air, warbles its few sweet notes, and then returns to almost exactly the same place again.

RICHARD'S PIPIT.

ANTHUS RICHARDI, Vieillot.

The only occurrence I find among published records is that in Byerley's "Fauna of Liverpool," where he says that the Rev. T. Staniforth informed him that he had a specimen stuffed that was killed at Crosby. In January, 1884, however, a Pipit was sent me for identification, which proved to be of this species, and which the sender, Mr. J. H. Wood, of Bury, said he had shot on the Wyre, not far from Fleetwood, in June or July, 1869. He remarks that "it was flying in and out of some gorse-bushes on the banks of the river Wyre, and I was struck by the peculiarity of its flight. It would fly out of one bush almost 'plumb' up into the air, and, after uttering a note something like a Sky-Lark, dart into the next thick bush, and remain for a few seconds."

ROCK-PIPIT.

Anthus obscurus (Latham).

The Rock-Pipit is a resident species, and is essentially a bird of the sea-coast, rarely, if ever, coming inland. It is thinly scattered along the whole length of the shore, but is not frequent anywhere. Byerley states that it occurs near Liverpool, and Mr. William Gillet of Preston shot several at Fleetwood and Cockerham in the winter of 1881, whilst Mr. W. E. Beckwith has many times observed it near Grange in autumn. Mr. Hugh P. Hornby informs me that he shot one at Pilling on January 26th, 1874, and Mr. J. B. Hodgkinson says that it breeds near Heysham, where he has seen nests, and that in winter it appears in the Ribble estuary. Mr. W. A. Durnford has seen it on Walney, and referring to this species on visiting that interesting island on May 30th, 1864, Mr. J. E. Harting writes (*Zool.*, p. 9163) that he found a nest, from which the bird rose, with four eggs, at the foot of a sand-hill where the long grass was growing rather more thickly. The eggs were much incubated.

FAMILY ORIOLIDÆ.—GENUS ORIOLUS.

GOLDEN ORIOLE.

Oriolus galbula, Linnæus.

An occasional visitor, of rare occurrence. In the *Mag. of Nat. Hist.* for 1829, on the authority of Mr. Blackwall, one (a female) is reported to have been shot, in July 1811, in Gorton fields near Manchester, and another at Quernmore Park near Lancaster, which latter is now preserved in the museum of that town. In 1868, according to the *Report of the Bury Nat. Hist. Soc.*, 1871, one was shot at Cockey Moor by Mr. C. Jackson, another being in its company, and Mr. John Ray Hardy some years ago, in the autumn, saw two specimens out of three or four which had been killed in one week in Hough End Clough, near Manchester. Mr. John Weld writes me that a pair was seen for a short time in the summer of 1870, frequenting the small gardens and enclosures in the bottoms near Chipping. The male bird was shot in a garden in Leagram, and the female was seen for a few days afterwards, but soon left the neighbourhood.

FAMILY LANIIDÆ.—GENUS LANIUS.

GREAT GREY SHRIKE.

LANIUS EXCUBITOR, Linnæus.

A rare autumn and winter visitor, which, however, has occurred too often to make its captures worth detailing; these including birds of the year, as well as males and females in full adult plumage. I have no records from north of the immediate neighbourhood of Preston, where one was seen by Mr. J. B. Hodgkinson in 1882 so early as the month of August. [The Rev. H. A. Macpherson informs me that he had records of a few examples from the Furness district.—Ed.] The latest date I have in spring is the 18th of April, when a bird was got in Heaton Park in 1867 by Mr. Wright Johnson. An interesting note by Mr. Blackwall, on the capture of one of this species by bird-lime, is quoted by Yarrell ("British Birds," 3rd edit., 1856) and runs as follows: " A birdstuffer following his occupation at Gorton, near Manchester, and having arranged the cage containing his call-birds, and placed his twigs well smeared with bird-lime in the manner best adapted to attain his object, he patiently waited the result. A Grey Shrike flew to the cage, most likely for the purpose of devouring the decoy-bird, and perching upon the limed twig attached to its summit, became entangled in the viscid material which covered it. The frightened Shrike made vigorous efforts to disengage itself from the unpleasant situation in which it was placed but without avail: its struggles only tended to involve it more completely in the tenacious toils with which it was encumbered. At length it was secured, and placed in a dark cage with the Redpolls which had been previously captured: but the surprise and mortification of the bird-catcher may be imagined, when on his arrival at home, he found that the Shrike had killed all its companions in captivity." Byerley also states ("Fauna of Liverpool," 1856) that one was caught some years ago striking at a noose set for Larks in the winter.

One shot at Shaw Hill, Chorley, March 20th, 1880; Lord Lilford's keeper shot one at Bank Hall, Bretherton, February 11th, 1890.—R. J. H.

RED-BACKED SHRIKE.

Lanius collurio, Linnæus.

A summer visitor; still breeding regularly, though in decreasing numbers, and stated by Blackwall to arrive

May 19th and depart September 13th. In the south of the county Byerley records it as breeding at Bootle, and Brockholes as at Rainhill (1860), and Mr. Peter Rylands in his "Catalogue" (*Naturalist*, 1837) gives it as occurring at Warrington, &c. Mr. J. B. Hodgkinson says that it bred at Farington and Penwortham twenty years ago, and that it is still found in some numbers, not being at all scarce in the year 1882, in which year also Mr. William Gillett saw both old and young at Midge Hall. At Lytham it has been observed by Mr. J. A. Jackson, and Mr. R. Standen has three times taken the nest, in May 1872, at Haighton, in June 1875, at Broughton, and in May 1878, at Goosnargh. Mr. H. Miller saw this Shrike early in May 1882, at Knott End, near Fleetwood, and was satisfied it was breeding, and has several times seen birds and eggs taken between Preston and Southport. In the Clitheroe district I only know of one occurrence, viz., on June 3rd, 1860, when Mr. J. J. Smithies took a nest (which was in a conspicuous place) and eggs from near Rimington; but on the other side of Pendle Hill, at Colne, Mr. T. Altham says that, up to a few years ago, the eggs were taken year after year in a brambly clough there. In Cliviger, Mr. H. Kerr states that he has seen this species once. Dr. Skaife writes (*Mag. Nat. Hist.*, 1838), that in the neighbourhood of Blackburn "this is by no means a rare bird, several specimens being procured every year," and in 1879 he told me that some years before, when walking in Bowland, he came across a thorn-bush, whose spikes were covered with beetles, &c. Mr. W. A. Durnford says that it was once common in Furness, and is still found in the Lake district. Mr. John Hardy tells me that a railway has cut up this Shrike's favourite spot on the south-west side of Manchester, and there is no doubt that such interferences, consequent on the great spread of population, are the main cause of its increasing rarity.

WOODCHAT.

Lanius pomeranus, Sparrman.

One has been killed near Lancaster, according to the MSS. of the late Rev. J. D. Banister, and Mr. F. Nicholson states (*Manchester City News*, 1875, on the authority of Mr. R. Entwistle of Bolton, who stuffed the bird) that one was shot at Walton-le-dale in 1870. These are the only records I have met with.

FAMILY AMPELIDÆ.—GENUS AMPELIS.

WAXWING.

Ampelis garrulus, Linnæus.

An occasional visitor in winter, sometimes appearing pretty numerously, and sometimes not being seen at all for several years. Many specimens were obtained in the winters of 1828-29, 1849-50, 1863-64, 1866-67, when large arrivals also took place all over the kingdom. The whole of the Lancashire examples were taken in the months of December and January.

FAMILY MUSCICAPIDÆ.—GENUS MUSCICAPA.

SPOTTED FLYCATCHER.

MUSCICAPA GRISOLA, Linnæus.

LOCAL NAME—*White Robin*.

A summer visitor; one of the last to arrive and first to leave, being seldom seen before the middle of May, or after the end of August. It may be called common everywhere, and in suitable localities becomes plentiful. Its nest is built oftenest in clefts of trees, and wedged between ivy or other climbing plants and the sides of walls, houses, and rocky banks, and four or five eggs are laid the end of May or beginning of June. If undisturbed, the same nesting-place is used for many years together.

PIED FLYCATCHER.

MUSCICAPA ATRICAPILLA, Linnæus.

A summer visitor, very local, and breeding in far fewer places now than was the case some score of years ago. It comes early, and its average date of arrival, according to Blackwall, is the 27th of April, and of departure the 4th of September. The same observer, in his "Researches in Zoology," writes: "On the 3rd of June, 1828, I procured a male Pied Flycatcher in the woods near the Ferry-house, on the western shore of Windermere, where I saw two males and a female. The female and one of the males had paired, and were occupied in constructing a nest in a hole in a decayed pollard ash on the margin of the lake." It is still found breeding in the same locality, near Hawkshead, but Mr. W. A. Durnford has only seen it once in Furness, and that at Dalton. In the Ribble valley it used to be found abundantly in Lord Ribblesdale's woods at Gisburne, just over the Yorkshire border, and Mr. W. Naylor informs me that a nest and eggs were taken at Whalley by Dr. Pinder in 1852. Near Clitheroe Mr. W. Peterkin has twice taken the nest, and the bird used to appear regularly in the woods of Waddow, but Mr. James Garnett has not seen one there since the 10th of May 1877. At Towneley, near Burnley, it used to breed every year, frequenting the gardens and neighbourhood of the brook; and at Frenchwood, near Preston, Mr. J. B. Hodgkinson says it was not uncommon some time ago. Mr. R. Standen states that the eggs were taken in June 1879, from a nest in an old bridge near Inglewhite, the only occurrence he knows of; and odd birds at various times have been seen or shot at Pilling, St. Michael's-on-Wyre, and Chipping: one at the last-named place in May 1882. Mr. R. J. Howard writes me that on the 9th of May 1883, he had a capital view of an adult male in Rufford Wood, getting within four or five yards of the bird, and was informed that there were three or four pairs breeding annually in the neighbourhood. A specimen was shot near Bury in June 1871, according to Mr. R. Davenport, and Mr. John Hardy says that it occurs in Trafford Park, near Manchester, as a regular summer visitant, and no doubt breeds there, although he has never seen the nest. If the hand of the collector and the amateur gunner could only be held for a while, I feel sure that the range of this pretty species would soon be largely increased.

FAMILY HIRUNDINIDÆ.—GENUS HIRUNDO.

SWALLOW.

Hirundo rustica, Linnæus.

A summer visitor, everywhere common. In mild seasons it will appear as early as the 10th or 11th of April, but is usually about a week later. There is a great want of unanimity in its arrangements for departure, the first flocks going before the end of August, and the bulk in September, whilst a few may be seen any year in October, and stragglers have over and over again been recorded early in November. In its breeding habits there is little variation, but Mr. J. F. Brockholes once saw a nest, composed of the ordinary materials, built in a tree at Maghull, and formed like a Chaffinch's among the twigs. The eggs are laid from early in June, and are generally five, but sometimes four, in number, two broods being hatched in the season.

[In *The Zoologist* for 1886 (p. 248), Mr. W. F. Brockholes has recorded the great destruction caused by the severe weather of the second week in May. As many as 150 Swallows and Martins were picked up at a country-house near Preston; nearly 100, almost all Swallows, were found at St. Michael's-on-Wyre, and 64 at a mill near Garstang; 392 Swallows, 65 House-Martins, 2 Sand-Martins, 8 Swifts, and 10 Land-Rails were found dead near Preston. This is merely one instance of a loss of life which was tolerably general in England.—Ed.]

GENUS CHELIDON.

MARTIN.

CHELIDON URBICA (Linnæus).

LOCAL NAME—*Martlet.*

A summer visitor; as common as the Swallow, but not arriving as a rule until a week or ten days after it, though leaving much about the same time. The Martin is subject to many vicissitudes in the course of its nest-building. Defects in the eaves and troughs under which it plasters its mud habitation often cause the nest to be washed down in heavy rains, and this frequently happens when the foundation is commenced too low, and a roof of a pent-house construction has to be made. It is probable that many of the late broods in autumn (occasionally as late as October) are those of pairs whose first homes have been destroyed in this way. Sometimes there are two holes in the nest. The Martin is double-brooded, and the eggs, generally four, sometimes three in number, are not laid before the beginning of June in most localities, though, apparently, a week earlier on some parts of the shores of Morecambe Bay.

Mr. R. J. Howard informs me that at Tarporley, Cheshire, where Sparrows are rigorously destroyed, he counted 32 occupied nests of House-Martins under 18 yards of eaves. I mention this, because the interference of the Sparrow with the House-Martin is general, and has an important bearing upon the decrease of this and other insectivorous species in Great Britain; also in America.—Ed.]

GENUS COTILE.

SAND-MARTIN.

COTILE RIPARIA (Linnæus).

A summer visitor, and one which is quite as common as the Swallow and the Martin, wherever there are suitable banks for it to bore its nesting-hole, though, owing to the absence of these in many localities, it is not as generally distributed. Odd birds are usually seen first of any of the Hirundinidæ, but the bulk arrive and depart at much the same periods as the Swallow. The favourite breeding-places of the Sand-Martin are the sandy banks of brooks and rivers, but Mr. W. A. Durnford says that it nests also on Walney Island, and it is often found in gravel-pits, and where land-slips have occurred, at some distance from any water. Fresh eggs may be found from the last week of May to the middle of June, and seven is the usual number, though many nests only contain six.

FAMILY CERTHIIDÆ.—GENUS CERTHIA.

TREE-CREEPER.

CERTHIA FAMILIARIS, Linnæus.

Resident, local, and nowhere common; having apparently decreased in numbers. In the Clitheroe district it was numerous in 1832, according to Mr.

Thomas Garnett (*Mag. Nat. Hist.*, 1832,) and he says of it, "not in winter so frequently observed as they otherwise would be, in consequence of their associating at that time with the different species of Titmouse, and using the same call-note, which is very different from that used by them when they are in single pairs, as is the case when they are not in company with the Titmice." Although both in the Ribble and Hodder valleys the Creeper breeds every year and may frequently be seen in winter, it is now far from "very common" as Mr. Garnett puts it. Mr. J. F. Brockholes in 1859 considered it more rare than formerly near Liverpool, and I have it reported as much scarcer near Urmston and Bury than it used to be. Mr. H. Miller finds it breeding in the woods about Clayton-le-Moors, and Mr. John Weld sees it the year through at Chipping. Near Goosnargh and St. Michael's-on-Wyre it has only been observed in winter, but the nest has been taken at Nateby, near Garstang, and there it occurs regularly, though not numerously. It is not common in the Preston district, and in Furness, according to W. B. K., Mr. W. A. Durnford's informant, it occurs but rarely. It is an interesting bird to watch when feeding: so rapid in its movements, so seemingly intent on the business in hand, no sooner has it wound its spiral track up the trunk of one tree than it takes its short flight to the bole of the next, putting it through the same process of examination, and occasionally following the course of a branch, though always clinging to the under side and hardly ever appearing on the upper. The nest is very often placed in trees overgrown with ivy, in a crevice between the ivy-stem and the trunk, and is very fully lined with feathers. The eggs are six in number, and are usually laid the latter half of April.

GENUS TICHODROMA.

WALL-CREEPER.

TICHODROMA MURARIA (Linnæus).

The only example of this continental species which has occurred in Britain since the first record, in 1792, was shot at Sabden, a village at the foot of Pendle Hill, on May 8th, 1872 (*Zool.* ss. p. 4839). It was seen flying about by itself, its crimson-banded wings drawing the attention of a lot of mill-hands, and was at length shot by a man named Edward Laycock, who took it to

Mr. W. Naylor of Whalley. Large slugs had been used to kill it, and it was so mangled that Mr. Naylor could not determine the sex, and had great difficulty in making it at all presentable. The specimen came into my possession.

[The above now forms part of the collection of Mr. J. Whitaker, of Rainworth Lodge, Notts. See "An Illustrated Manual of British Birds," p. 111, for a concise account of this species; to which I may add that the Wall-Creeper is now known to wander, not infrequently, to Normandy.—Ed.]

FAMILY FRINGILLIDÆ.—SUBFAMILY FRINGILLINÆ.
GENUS CARDUELIS.

GOLDFINCH.

CARDUELIS ELEGANS, Stephens.

LOCAL NAMES—*Flinch, Redcap.*

Resident, but so decreased in numbers as to be almost extinct. The march of agriculture is one great reason for this; waste lands where thistles (the seed of which, especially of *Centaurea nigra*, L., is its favourite food), groundsel and nettles used to grow in plenty, being now so largely brought under cultivation. The bird-catcher too (or as he is more commonly called the "tuttler" or "touter," *i.e.* one who entices), is the deadly enemy of the Goldfinch, and any stray individuals are at once captured to satisfy the exigencies of the demand from the large towns. I have few records of any nests for the last twenty years, but before that time Mr. W. Peterkin says the Goldfinch used to breed regularly near Clitheroe, and was not at all scarce, while Mr. J. B.

Hodgkinson avers that there used to be a nest in every orchard near Grange, and that in the Preston district also there were plenty of birds; he saw four or five among some thistles near the latter place on the 5th of October, 1882. About Liverpool Mr. J. F. Brockholes (1859) recorded it as scarce, but that he had several times seen the nest, in sycamore-trees, near the extremity of a slender branch. Mr. John Weld says that it breeds at Chipping, visiting the place occasionally, and that a flock of twenty or more was seen there feeding on dock seeds on the 16th of December, 1882. Mr. R. Standen sometimes sees small flocks among the thistles in autumn near Goosnargh. Mr. John Hardy informs me that in the neighbourhood of Manchester it is not nearly so common as it was twenty-five years ago, either near the city or in more retired places, but it occurs still, and nests regularly in orchards and gardens. It is very rare now in the Wyre valley, and Mr. Hugh P. Hornby has only once seen a bird of late, namely in August 1874, near Poulton-le-Fylde: in the same year, in May, Mr. J. E. Palmer writes me that he saw a pair near Todmorden, and thought it probable they were nesting. In Furness the Goldfinch is reported by Mr. W. A. Durnford to be resident in small numbers, and Dr. C. A. Parker says that it still breeds on the Cumberland border, though getting rarer and rarer.

"'In the spring of 1887 four young Goldfinches were taken from a nest in the garden at Pleasington Hall, within two miles of Blackburn, and one of these birds was living on January 3rd, 1889. This is the most recent instance I have of the breeding of the Goldfinch in this neighbourhood. Every winter small flocks are seen in the Ribble valley."—R. J. H.[1]

GENUS CHRYSOMITRIS.

SISKIN.

CHRYSOMITRIS SPINUS (Linnæus).

A winter visitor: the only instance in which it is reported to have bred being that recorded by Yarrell (3rd edit.), who says: "Mr. Howitt of Lancaster sent me word that large flocks containing several hundred birds have been seen there during the winter; a few remained in the summer of 1836 to breed; six pair of old birds were seen about, and later in the season several young ones." It appears also to be now much less plentiful in winter than it was twenty years ago, and the same statement is made from all parts of the county—that it is occasionally seen and shot, but that it is of increasing rarity. It is most usually seen consorting with flocks of the Lesser Redpoll.

["Seen each winter in Livesey and Balderston, in small flocks, seldom exceeding 20, feeding on seeds of the alder."—R. J. H.]

GENUS LIGURINUS.

GREENFINCH.

LIGURINUS CHLORIS (Linnæus).

LOCAL NAMES—*Green Linnet, Greenbull.*

Resident and common, breeding numerously throughout the county. In the more exposed districts it is migratory in winter, retiring to sheltered spots, and the severe seasons of 1878, &c., thinned its numbers very much everywhere. In 1875 Mr. John Plant says that a pair bred in Peel Park, Salford, and altogether it is very partial to the neighbourhood of houses and gardens, nesting in the hedges surrounding the latter, and levying heavy black-mail on newly-sown turnip, radish, and other seeds. Mr. C. E. Reade remarks that the Greenfinch is the only bird he has noticed take the seed of the Mezereon (*Daphne mezereum*). Mr. John Hardy alludes to its curious manner of singing during flight, suspended, as he says, in somewhat the manner of the Wood-Lark, but low down and not so long continued. It lays its six eggs from the middle of May to the beginning of June.

["Scores of Greenfinches frequent Eanam Brewery yard, in the centre of Blackburn, during hard weather, to feed on the seed of the spent hops."—R. J. H.]

GENUS COCCOTHRAUSTES.

HAWFINCH.

COCCOTHRAUSTES VULGARIS, Pallas.

A resident species, very local, and more generally distributed in winter, but which of late years has extended its breeding-range northwards, and has been found nesting in several localities where previously it had not been observed in summer. Mr. John Weld informs me that a pair were observed in the kitchen garden of Leagram Hall in July 1878, and seemed to have bred somewhere near, two young birds, unable to fly, being captured close by. The young were placed in a cage, and ate with good appetite the peas which were given them, and after release, the family together committed serious ravages among the growing pods. No one in the neighbourhood had previously seen the species, but it was again noticed in April 1879, July 1880, and April 1881. In the summer of 1880 Mr. W. Fitzherbert Brockholes tells me a young bird was picked up dead near Claughton, and Mr. John Watson reports that, in the same year, a nest with three eggs (the female being caught) was taken near Coniston, it otherwise being only known as a rare winter visitor there. Mr. W. Gillett saw six birds in the breeding-season of 1881 near Chorley, and a young male is still in existence which was rescued from a nest of young with which some children were playing near Whalley in that year. The owner of this specimen said that a year or two before, he had seen a lot of young which, from their size, must have been bred in the neighbouring woods, by the side of the Calder. Mr. R. Standen says that in the autumn of 1882 about a dozen birds, two of which (males) he saw, out of four shot, appeared in a garden near Goosnargh, and were feeding on old plum-stones; and on the 2nd or 3rd of the previous May a male, in full breeding plumage and good condition, was sent to Mr. John Hardy, from Worden Hall, near Preston. This last was picked up, scarcely alive, after a severe storm, and soon died. Mr. Hardy says the Hawfinch is not a very infrequent visitor near Manchester, it having been several times shot both in summer and winter, and that in two cases he has known of its nesting in the neighbourhood. In the spring of 1883 Mr. J. J. Hornby found it breeding at Knowsley, and it again appeared in most of the localities before mentioned, whilst during the winter, on more than one estate, it was observed in considerable flocks, fifteen to twenty-five birds being counted in one tree, but being very shy and wary. Should this handsome bird continue to show a disposition to breed, it is sincerely to be hoped that it may remain unmolested.

[According to notes recently furnished by Mr. R. J. Howard and Mr. W. F. Brockholes, the Hawfinch is still increasing generally. It may be met with in considerable numbers near Chorley from December to the end of March; while many instances of nesting are recorded. The following from Mr. Howard deserves quotation in full:—" The irides, in a live bird, are not greyish-white but vinaceous. On August 7th, 1884, my male Merlin dashed from his bath at the Hawfinch which Billington brought from Redcar. I picked the bird from the cage-floor, as it was dying, and paid particular attention to the colour of the irides; the bright madder-brown got gradually lighter, until at last—before the bird was cold—it had faded away, and the colour could only be described as greyish-white. I had often held the bird in my hand, so that I could closely examine the eyes, and found the colour arranged in concentric circles, those near the pupil being brightest; the intensity of colour varied when I teased the bird."—Ed.]

GENUS PASSER.

HOUSE-SPARROW.

PASSER DOMESTICUS (Linnæus).

Resident, and exceedingly abundant everywhere. Its nest is most commonly fixed in holes in buildings, but, in the Fylde especially and near Southport, large numbers are built in trees among the branches, having the entrance-hole sometimes at the top and sometimes at the side. The eggs are very varied in their markings and are from four to six in number, being laid from the end of April to July. As many as three broods will be reared in a season.

TREE-SPARROW.

Passer montanus (Linnæus).

A resident species, but very local. It is probably commoner in the neighbourhood of Manchester than anywhere else; breeding there in holes in the trunks of trees, and occupying the same hole year after year, even when buildings have begun to spread. I transcribe some interesting notes furnished to the *Naturalist* of 1865 by Mr. J. Chappell: he says, "The Tree-Sparrow occurs in the neighbourhood of Manchester, generally building in holes in decayed willow, poplar, and oak trees, near the banks of streams. I have known as many as four or five nests in one tree, and sometimes a nest of the Starling in the same. This season I have noticed the following strange occurrence. A tree having been cut down in which some Tree-Sparrows have been in the habit of building for the last twenty years, a pair have adapted themselves to circumstances, and built a large oval nest in a hawthorn hedge, about twenty yards from the place where the tree stood; it contained three eggs. I found an old nest of the same description in the same hedge a few yards nearer the tree. I waited the return of the birds lest I might confound it with the House-Sparrow, the nest of which it [sic] greatly resembled: after about an hour's waiting my patience was rewarded by seeing the birds, and one of them entered the nest, so that I was satisfied I had made no mistake." The Tree-Sparrow has been noted as occurring on Chat Moss by several observers from Blackwall downwards, and Mr. C. S. Gregson (*Ibis*, 1865, A. G. More) says he has taken the nest both near Warrington and Lancaster. Mr. John Weld has seen several specimens which have been shot near Chipping this last two or three years, but has not discovered their breeding-place, and Lord Lilford informs me he has several times seen it on the Bank Hall estate, near Atherton. Mr. J. B. Hodgkinson says that it still breeds within a short distance of Preston, and Mr. William Garnett is sure it nested in the Hodder valley, near Bashall, shortly before 1880. In winter it is more general, and in hard weather mixes in considerable flocks with other species. Its habits are, however, rather solitary, and Mr. John Hardy says that he has often watched it in winter about a farm with the House-Sparrow and other birds during the day, and at nightfall has seen it leave them, and roost singly among the ivy and twigs growing on the upper parts of the trunks of trees in cloughs and small woods. The eggs are smaller than those of the common species, and more uniform in colouring and shape; the note also is different, being more harsh.

GENUS FRINGILLA.

CHAFFINCH.

FRINGILLA CŒLEBS, Linnæus.

LOCAL NAMES—*Pink*, *Spink*, *Bullspink*, *Flackie* or *Fleckie* (from its flecked wings).

Resident, and everywhere common; breeding not infrequently in the immediate vicinity of the large towns. In Peel Park, Salford, it was once well known, and there it used to construct its nest very largely of the scraps of raw cotton blown about from the neighbouring manufactories. It is a hardy bird, capable of standing very severe weather, and the winter flocks are composed almost entirely of males; females, though uncommonly, being sometimes with them in small numbers. It is viewed with considerable mistrust by the country people when seen about their gardens, and this feeling, at least near Whalley, finds expression in the couplet,

"The Spink and the Sparrow
Are the devil's bow and arrow,"

the sentiment generally following the more widely-used one,

"The Robin and the Wren
Are God's cock and hen."

The Chaffinch does not sing in winter, but begins very early in spring, and in February a sunny day will generally set it going. When disturbed, either at its nest or at any other time, it is very noisy, uttering its alarm-note incessantly, and these cries of distress are echoed by all its feathered neighbours. Mr. Thomas Garnett (*Mag. Nat. Hist.*, 1822), in remarking on birds understanding each others' notes of woe, relates how he made a fledged young Thrush cry out, and how, following the parents' alarm-shriek, he heard the Blackbird, Chaffinch, Tit-Lark, Redbreast, Oxeye, Blue and Marsh-Tits, Wren, and Goldcrest; the Creeper alone seeming neither to understand nor care. The Chaffinch lays its five eggs from early May to early June.

BRAMBLING.

FRINGILLA MONTIFRINGILLA, Linnæus.

A winter visitor, irregular in its appearance, and very much more numerous some years than others. The stubble-fields of the Fylde and other similar localities along the coast are its favourite haunt, and it is only occasionally noticed in the more inland districts. Mr. Hugh P. Hornby says that before the hard winters ending with that of 1880 it used to appear in large flocks near St. Michael's-on-Wyre, but that since then it has not been seen. On the Formby shore Mr. C. S. Gregson has found it very plentiful some winters, usually moving about a good deal, and not staying long at one place. It remains from the end of October to the beginning of April.

GENUS LINOTA.

LINNET.

Linota cannabina (Linnæus).

Local Names—*Grey Linnet, Brown Linnet, Redcap, Whinfinch, Gorse-cock, Paywee.*

Resident, partially migratory in winter, and moving about at that season in large flocks over the open country in search of food. It is most plentiful on the coast, and from the shores of Walney to the neighbourhood of Liverpool it breeds commonly in the furze and low stick-hedges which characterize this section of the county. In many inland districts, too, it is numerous, affecting there the low mosses, or the plantations of whin-bushes which skirt the bases of the moorlands. Mr. C. E. Reade says it breeds plentifully on the mosses near Urmston, and according to the Report of the Bury Nat. Hist. Society (1871) it is found the year through in that district, being very common in the stubble-fields in winter. Mr. John Hardy informs me that it is universally distributed in the Manchester neighbourhood, breeding in hedges—or if in furze-bushes, generally near hedges—and at that time of the year being quite solitary. On the moors from Rochdale through Bacup to Cliviger it still occurs, though in diminished numbers. Dr. Skaife wrote of it in 1838 (*Mag. of Nat. Hist.*, 1838) as very common near Blackburn, but Mr. R. J. Howard only considers it moderately so now, and it is curious that it is entirely absent from the Clitheroe district at all seasons. It is common near Chipping, Goosnargh, and St. Michael's-on-Wyre, and at the last-named place Mr. Hugh P. Hornby says it roosts in winter in the rhododendron-bushes in large flocks. Mr. Henry Miller has found this species moderately distributed over the Fylde, and thinks its numbers a little increased lately, though far fewer now altogether than was the case twenty years ago : the raids of the bird-catchers have much to do with this greater rarity. In Furness Mr. W. A. Durnford says it is far more numerous at the migratory season than any other, but is always resident. The large flocks which are seen in winter are almost entirely composed of Linnets, though sometimes with a slight admixture of Twites and other allied species, and Mr. Hardy remarks that they are all very partial to the seeds of various species of *Polygonum*, especially *P. aviculare*, but the seeds of some *Labiates* (*Stachys* for example) will cause them to visit one place frequently, even in the face of danger. The eggs are laid late in April or early in May, and are four to six in number, generally five.

LESSER REDPOLL.

Linota rufescens (Vieillot).

Local Names—*Jittie, Grey Bob, Chivvy.*

Resident, and breeding regularly ; but, like the Linnet, more generally distributed in winter, and being seen then in small flocks almost everywhere, usually unmixed with other species. It breeds in small quantities in most parts of the southern division of the county, becoming more numerous going north, and in the valley of the Ribble from Preston upwards, and also in that of the Hodder it is plentiful. Further north still, in the district watered by the Wyre, it is rarer, and in Furness Mr. W. A. Durnford has not observed it personally, though he believes it occurs at times. High hedges, and the forks of fruit- and other trees are the favourite situations for its nest, which is usually a very neat and tidy structure, though not invariably so, as I have seen one now and then built in a most slovenly manner. The birds are exceedingly solicitous if the nest be approached, and hop about, plaintively chirping, within less than arm's length of the intruder. The Lesser Redpoll brings up two lots of young; the first eggs, which number five or six, being laid the third week in May, and the construction of the second nest being commenced before the young of the first have flown.*

TWITE.

Linota flavirostris (Linnæus).

Local Names—*Manx Linnet, Tricefinch.*

Resident; occurring on open moorlands, whether at a high or low level, and breeding as commonly on the South Lancashire mosses as in more elevated districts. On the moors of the eastern border it is numerous, but near Colne it is now much rarer than it used to be, and on Pendle Hill it only breeds occasionally. In 1838 Dr.

* [Mealy Redpoll, *Linota linaria* (Linn.). Mr. R. J. Howard writes:—"I am convinced that this bird was occasionally caught on Mellor Moor, near Blackburn, about twenty years ago. Joseph Ward, our wood-bailiff, who takes much interest in cage-birds, says that when catching Lesser Redpolls he got a few larger and greyer birds with white wing-bars. He at once recognized the bird when I showed him a skin of the Mealy Redpoll in winter plumage."—Ed.]

Skaife considered it plentiful near Blackburn, and Mr. R. J. Howard informs me that about the year 1874 Mr. W. L. Constantine found upwards of twenty nests in the vicinity of the Roman beacon on Mellor Moor: it seldom, however, appears there in such numbers as this, and is usually much more abundant in autumn and winter than summer. Mr. J. B. Hodgkinson says that it still breeds on the mosses near Preston, as also on Longridge and Beaton Fells: but Mr. W. A. Durnford has not personally identified it in Furness, though he says he has good authority for inserting it in his list of the birds of that district. It leaves the higher grounds in winter, and approaches the towns, feeding, in company with its congeners, in considerable flocks on the stubble-fields and waste lands. Mr. John Hardy considers it the most common species of the genus near Manchester, out of the breeding-season, and thinks the supply from the immediate neighbourhood sufficient to account for the size of the flocks, without their being, like those of the Linnet, increased by individuals from other and more distant districts. He writes:—" During the years of the Cotton Famine, when the factory girls wandered in their neighbourhoods further away than is usual with them, knitting or otherwise employing themselves to kill the time, I found some nests of the Twite lined with lengths or bits of worsted, one in particular being lined with white in a neat and remarkable manner." This bird is more or less gregarious in the breeding-season, several nests being usually within a small area: Mr. T. Altham tells me he once found a nest on Pendle on which the old bird sat singing. Five eggs are laid towards the end of May.

SUBFAMILY LOXIINÆ.—GENUS PYRRHULA.

BULLFINCH.

Pyrrhula europæa, Vieillot.

Local Names—*Thick-bill, Nope.*

Resident, and very evenly distributed; nowhere common as a breeding species, there being seldom above two or three pairs in any one district. It is a bird of shy and retiring habits, nesting usually in thick bushes, though Mr. T. Altham has found a nest on a spruce-branch, and Mr. J. B. Hodgkinson has seen one near the Winster on yews. On the Cumberland border Dr. Parker says the Bullfinch is mostly seen in winter, and it is perhaps most commonly observed at this season in other localities, being often disturbed by parties of cover-shooters. It lays five or six eggs about the end of May.

GENUS PINICOLA.

PINE-GROSBEAK.

Pinicola enucleator (Linnæus).

The Rev. H. Clark states (*Zool.*, 1845, p. 1025) that he had a Pine-Grosbeak which was killed in February, 1845, in a fir-plantation near Rochdale, and Mr. Peter Rylands in his "Catalogue of Birds found in Lancashire" (Neville Wood's *Naturalist*, 1837) gives "Hulston fir-trees, T. K. G." (Glazebrook, F.L.S.) as another locality where the species had occurred. Mr. J. H. Gurney, jun. (*Zool.*, 1877, p. 242 *sqq.*) doubts the locality of the first instance, and the identity of the second, but I find no adequate reason for his conclusions. Mr. Rylands, at present [1885] the member for Burnley, has written me that "Hurlston" is the correct spelling, and this appears to be in the neighbourhood of Ormskirk.

GENUS LOXIA.

CROSSBILL.

Loxia curvirostra, Linnæus.

An autumn and winter visitor, occurring, usually in flocks, at irregular intervals. Reasonable grounds exist for supposing that it occasionally remains to breed, but no specific evidence is available, and for the present the point must remain unsettled. The Crossbill has been seen and shot in most parts of the county between July and March. Mr. W. Eagle Clarke writes me that on August 2nd, 1883, one was seen in a garden near Morecambe by Mr. John Grassham; it was in greyish-green plumage, and he took it to be a bird of the year; it was also very tame, permitting approach within a few yards.

SUBFAMILY EMBERIZINÆ.—GENUS EMBERIZA.

CORN-BUNTING.

EMBERIZA MILIARIA, Linnæus.

LOCAL NAMES—*Bunting, Grey Bunting.*

Resident, migratory in some districts, and moving about in winter in company with Larks and Yellowhammers. It is very locally distributed, and being seldom seen except where grain is grown, it is commonest on the flat lands in the southern half of the county. It is not so plentiful on the ploughed country drained by the Ribble and Wyre, and Mr. R. J. Howard says that of late years it has become much less numerous in the neighbourhood of Rufford and Tarleton, where formerly it was not uncommon. It breeds sparingly near Preston, and nests have been taken at Newsham and Cockerham, but in the Fylde generally it is rarely met with. With the conversion, in the north-east, of most of the arable into grazing land, it has disappeared during the last twenty years from the Blackburn district, where Mr. W. L. Constantine used to find it frequenting Revidge; and higher up the Ribble it is not known. Mr. W. A. Durnford writes of it as resident in small numbers in Furness, and Dr. Parker considers it fairly numerous in the portion adjoining the county of Cumberland, whilst nearer the Winster, Mr. J. Watson says it is common. According to Mr. John Hardy, it is a late breeder, and its nests, of which he has seen more during the last thirty years than of any other of the Buntings, are placed further into the fields. Though so inconspicuous in plumage, its habit of uttering its short warble from the top of some bush or stone wall makes its presence very noticeable.

YELLOW-HAMMER.

Emberiza citrinella, Linnæus.

Local Names—*Bessie, Goldfinch, Goldenfinch, Yellow Ring, Yellow Urin or Yuring, Yellow Yoldring.*

Resident and common, breeding numerously everywhere, except in the comparatively bare district about Accrington and Bacup. In winter it is partially migratory, and traverses the country in small flocks in search of food. Mr. R. Standen says it is much rarer than it used to be about Goosnargh, neither does he see it in winter so often, and Mr. Hugh P. Hornby makes the same remark as to the neighbourhood of St. Michael's-on-Wyre; but with these exceptions it appears to be increasing its numbers. It begins its song very early in the year, if encouraged by a day or two's fine weather, and I have heard it warbling "a little bit o' bread, no chee—se" before February was out. It lays four or five eggs, often only three, the first fortnight in May, and the long grass in hedge-bottoms is the favourite situation for its nest. Mr. James Murton reports (*Zool.*, 1871) his having found a nest with three young nearly fledged on the 5th of September, an extraordinarily late date.

CIRL BUNTING.

EMBERIZA CIRLUS, Linnæus.

Mr. C. E. Reade states that at Urmston, fifteen or twenty years ago, the Cirl Bunting was occasionally found in winter with the flocks of Yellow-hammers and Finches, but since that time it has not been observed. Mr. C. S. Gregson writes me that it has bred in his warren at Formby, and that two eggs and one bird from there are now in his collection. Elsewhere there are no records.

ORTOLAN BUNTING.

EMBERIZA HORTULANA, Linnæus.

A fine male, killed near Manchester in November 1827, was recorded in April 1828 in the *Zoological Journal* (iii. p. 498) by Yarrell, and after passing into his possession it was figured by Selby.

REED-BUNTING.

EMBERIZA SCHŒNICLUS, Linnæus.

LOCAL NAMES—*Black-headed Bunting or Bodkin, Black-cap, Reed-Sparrow, Pit-Sparrow.*

Resident and common, though irregularly distributed, and especially so in winter, when it leaves some districts altogether. Near Clitheroe, for instance, it is an absolute migrant, disappearing in October, and usually coming again early in April, though sometimes a month sooner; always the first of the spring arrivals, and the males often before the females. Mr. C. E. Reade says that at Urmston, in the south of the county, he sees it more frequently in winter than at any other time, and that it remains through the hardest frost; being solitary in its habits, and not more than one or two being generally seen together. In summer it is found everywhere in suitable localities, nesting oftenest in wet places near pits and ponds, but not seldom among the long grass of young plantations. Its four or five eggs are laid the first fortnight in May.

GENUS CALCARIUS.

LAPLAND BUNTING.

CALCARIUS LAPPONICUS (Linnæus).

A rare winter visitor, of which I find only the following occurrences noted. One (*Mag. Nat. Hist.*, 1834), supposed to be a young male, shot near Preston, bought October 18th, 1833, in the Manchester game market, and afterwards finding a place in the Museum of the Natural History Society of that city. A young male, purchased alive in the Liverpool game market, from among a cageful of Sky-larks, by Mr. N. Cooke (*Zool.*, 1867) on the 27th of October, 1866, and kept in his aviary for some time; it had been captured along with the Larks on the Southport sand-hills. One, supposed to be a young bird, shot in November 1865 on Whitemoss, near Middleton, and now in the collection of Mr. James Holland. One caught alive on the Formby coast in the winter of 1881-82, and kept alive for some time by Mr. C. S. Gregson, who informs me that it was a young bird, and in the spring of 1882 moulted into male plumage.

GENUS PLECTROPHENAX.

SNOW-BUNTING.

PLECTROPHENAX NIVALIS (Linnæus).

LOCAL NAMES—*Shore-Lark, Sea-Lark, Snowbird, Snowflake.*

A winter visitor, occurring in flocks most commonly along the coast, and frequenting many inland localities as well, both the tops of the highest hills, and waste places on the lower grounds. In ordinary weather not more than a score are usually seen together, but in strong frosts the flocks often contain as many as a hundred birds, which then consort with Larks, Linnets, and other species of like habits. Mr. T. Jackson informs me that near Sunderland point they come as soon as the ground gets covered with snow, a few at first, and then hundreds if the weather continues frosty; their favourite food being the Glasswort (*Salicornia*) which grows plentifully there on the marshes, and is bared by the receding tide. He writes: "they are so tame from hunger that you could almost knock them down with a stick, and as soon as the frost is over, we never see more of them." Mr. John Weld says that large flocks are found most winters on the tops of the hills in Chipping and Bleasdale, after the first heavy fall of snow, and that on November 18th 1878, a band of at least a hundred was seen on Parlick. In the severe winter of 1879 Mr. H. Miller saw very many on Haslingden Moors and Hapton Scouts; they being so tame that they came to pick the remnants from the horse-feed as the carts were being loaded with stone at the quarries. Mr. John Hardy says they occur in varying quantities every winter near Manchester, and that they were pretty plentiful even in that of 1882, which was so mild [in England, but not on the Continent.—Ed.]. He once, thirty years ago, kept in confinement a number of birds, which he had procured alive out of a flock which frequented the neighbourhood of Chorlton from December to March, and they seemed so healthy that he had hopes of their breeding, but early in June they became very restless, and all died off without any apparent cause. The flocks mostly, though not always entirely, consist of birds of the year, and never appear before November; stragglers being occasionally seen as late as the middle of April.

FAMILY STURNIDÆ.—GENUS STURNUS.

STARLING.

STURNUS VULGARIS, Linnæus.

LOCAL NAMES—*Stare, Shepster, Sheppie.*

Resident, and everywhere common, leaving, however, the more exposed districts in winter. At this season it collects in small flocks, which, throughout the day, in company with Rooks generally, seek for food in the pastures, and towards evening join together, forming immense multitudes, and passing the night in sheltered situations, as among thick trees, &c. When freshly alighted at such places, the noise made by the calls of the birds to each other passes all belief. The Starling is a largely increasing species, and is so prevalent that I think it probable even these enormous flocks may be formed locally. Its capacity of adapting its nest to many varieties of situation, its recognized harmlessness to the agriculturist, and its own natural sharpness, give it great chances of increase, and of these it fully avails itself. Perhaps the most curious place chosen is that recorded in the *Field* of October 31st, 1874, by Mr. John Wrigley of Formby, who says he found a nest with young ones in it, built upon the ribs of a wrecked ship, which at high tide was not less than a quarter of a mile from the shore. Usually the nest is in a hole either of tree, cliff, or wall, but it is often fixed among ivy, and Mr. T. Altham has known this bird many times to build a big loose nest on the branches of the spruce-fir. [On May 5th 1884, Mr. R. J. Howard took four fresh Starling's eggs from a Magpie's nest in a plantation near Rishton Reservoir.] It is an incomparable mimic, and will sometimes deceive the most trained ear; its imitation of the Curlew is almost perfect. The Starling usually breeds in colonies, laying five to seven eggs late in April, and although the contrary has been asserted, does not, I believe, rear more than one brood. [This has been again questioned.—Ed.] It may often be seen examining the previous year's nesting holes as early as the month of February if a fine day or two happen to occur.

GENUS PASTOR.

ROSE-COLOURED STARLING.

Pastor roseus (Linnæus).

An occasional visitor in autumn, of rare occurrence. Latham ("Gen. Hist. of Birds," 1821) wrote that it was said one or more Pastors had been killed, almost every season, about Ormskirk, and Blackwall has reported particulars of two specimens which were in the Manchester Museum: one (*May. Nat. Hist.*, 1829) shot many years before in Salford, and another (*id. op.*, 1831) killed near Eccles on August 19th, 1830. This last was a male, and the contents of its gizzard were found to consist principally of the larvæ of insects, the indigestible parts of beetles, and a few seeds of vegetables. Byerley records an instance of one being killed near Liverpool about the year 1840, and Mr. Anthony Mason, of Grange, informs me one was shot at Cartmel whilst feeding with Starlings in the autumn of 1854. A specimen was taken near Oldham in 1860, and Mr. David Mitchell of Lowerhouse had one which was killed near Lancaster, where, he said, examples had occurred in four successive years. Mr. John Hardy has known of three instances of its occurrence during the last thirty-three years: one in Prestwich Clough, now in the museum of Owens College; another, a young bird, shot in or near Heaton Park; and the third killed near Barlow Wood, on the south of Manchester.

["One was shot near the windmill, Ainsdale, and purchased from Mr. Riddiough by the late T. Eccleston, Esq., Scarisbrick." (Glazebrook's "Guide to Southport," p. 150.)—Ed.]

FAMILY CORVIDÆ.—GENUS PYRRHOCORAX.

CHOUGH.

PYRRHOCORAX GRACULUS (Linnæus).

Occasionally, but very rarely, seen on the coast; generally after heavy weather, and probably storm-driven from the Isle of Man or North Wales.

GENUS GARRULUS.

JAY.

GARRULUS GLANDARIUS (Linnæus).

LOCAL NAME—*Jaypiet.*

Resident, but every year decreasing in numbers; its propensity for emptying Pheasants' eggs causing it to be mercilessly destroyed by the game-keeping fraternity. It is very shy and wary, and seldom falls to the gun; poison being the chief element in its extermination. It frequents large woods, especially those at high level, choosing the most secluded portions, and in such situations is found throughout the whole of the county from north to south. It lays five to seven eggs the end of April or beginning of May.

GENUS PICA.

MAGPIE.

PICA RUSTICA (Scopoli).

LOCAL NAMES—*Maggie, Piet, Pienot, Pianot, Pie-Annet.*

Resident, and breeding in all districts; its abundance anywhere being in exact inverse proportion to the vigour with which it is pursued by the game-keepers. There is no doubt of its sins as an egg-stealer, but the Rook is quite as dangerous in this respect, and the enmity which the Magpie excites amongst sportsmen is perhaps a little unreasoning. The farmer, too, often complains of its raids on the eggs of his stray-laying hens; but in Norway, where its presence is so encouraged and where almost every farm-yard has its nest, very large quantities of poultry are kept, and the Magpie is evidently supposed to provide compensation in some way. In Lancashire it is found breeding mostly in the secluded woods at the foot of the fells, and is still fairly common in wooded districts, regularly building its nest, if unmolested, in close proximity to dwellings. In winter, three or four individuals may generally be seen together in the open country, and on the sea-shore, where the land is manured with shell-fish, the flocks often number two or three score. From five to seven eggs are laid late in April or early in May.

GENUS CORVUS.

JACKDAW.

CORVUS MONEDULA, Linnæus.

Resident, and distributed in small numbers over the whole of the county. It is reported as increasing in most places, but I do not find this to be the case in my own district of Clitheroe, where its numbers appear to me to remain pretty constant. A few pairs breed in the church steeples of almost every town and village, though in Manchester it has now deserted many of those which formerly it regularly occupied. Holes in trees, chimneys, and walls of ruins are favourite nesting-places, and its senses are remarkably acute, an approaching footstep being immediately detected, and the nest left, though it be so much as six or seven feet from the outer air and thirty feet from the ground. It affects very much the company of Rooks, often breeding in the immediate neighbourhood of Rookeries, and a few generally being mixed with the winter flocks. The eggs are laid from the end of April to the beginning of May, and are five or six in number.

CARRION-CROW.

CORVUS CORONE, Linnæus.

LOCAL NAMES—*Craw*, *Crow*, *Carr-Crow*, *Doup*, *Doup-Crow*, *Ket-Crow*. (*Ket* = old flesh or filth.)

Resident, but getting scarcer every year, especially where there is any game-preserving; the nests being riddled with shot, while the birds are tempted with all sorts of dainties, merely seasoned with a little strychnine, laid conspicuously in the places they frequent. Persecution, however, has made them so wary, that they are exceedingly difficult to get rid of; while they are so persevering that they will build two or even three nests in a season, if the first be destroyed, and fresh eggs are regularly found in June. Being wild and solitary in their habits, Crows oftenest fix their nest in retired woods or in the sparsely-scattered trees of unfrequented cloughs and gullies, and there it is seldom at a great height from the ground: indeed, among the hills of the Lake district it is placed in thorn-bushes, hardly out of arm's length. A situation from which there is a good look-out is evidently the first requisite, and though high trees are generally chosen for the nest in the vicinity of a farm or dwelling-house, I have seen one year after year, in the very heart of the strictest game-preserve I know, on quite a low tree, and with only the advantage of a good view all round. The nests are very thickly lined with sheep's-wool, and the eggs are laid early in April, the number being five or six.

HOODED CROW.

Corvus cornix, Linnæus.

Local Name—*Manx Crow*.

A winter visitor, only seen as a straggler inland, and more commonly noticed on the coast, though it is much rarer there than it used to be. It is seldom seen before October, Blackwall's dates of arrival and departure being October 30 and April 13, but Mr. John Weld saw seven birds among Rooks in the meadows of Leagram Hall in 1876 on August 15, an early date. It was very common on Martin Mere, near Southport, before it was so entirely reclaimed, and John Cookson, an old fowler there, told Mr. R. J. Howard that it was a perfect nuisance, for if a duck were winged or killed, and not recovered before night, the Hoodies would find it at daybreak and pick it clean. Of Martin Mere, Baines ("Hist. of Lancashire," 1868-70, vol. ii. p. 431) says it "originally comprised 3,132 statute acres of land," and Leland ("Itinerary," Oxford, 1770, vol. vii. p. 49) nearly 300 years ago described it as "the greatest meare of Lancastreshire a iiii. miles in Lengthe and a iii. in Bredthe." In 1693 Thomas Fleetwood of Bank Hall began to drain it, and in 1849-50 Sir Thomas Dalrymple Hesketh of Rufford drained his portion, and brought 800 acres of land into cultivation. Parts of it are still swampy, and are occasionally flooded, but no portion is constantly under water. Dr. Leigh in his "Nat. Hist. of Lancashire" (pp. 158-9), published in 1700, says of the Sea-Crow,* which I take to be the species under

---

* Willughby ("Ornithology," ed. Ray, 1678) says that "Mr. Ralph Johnson calls the Royston (Hooded) Crow the Sea-Crow."

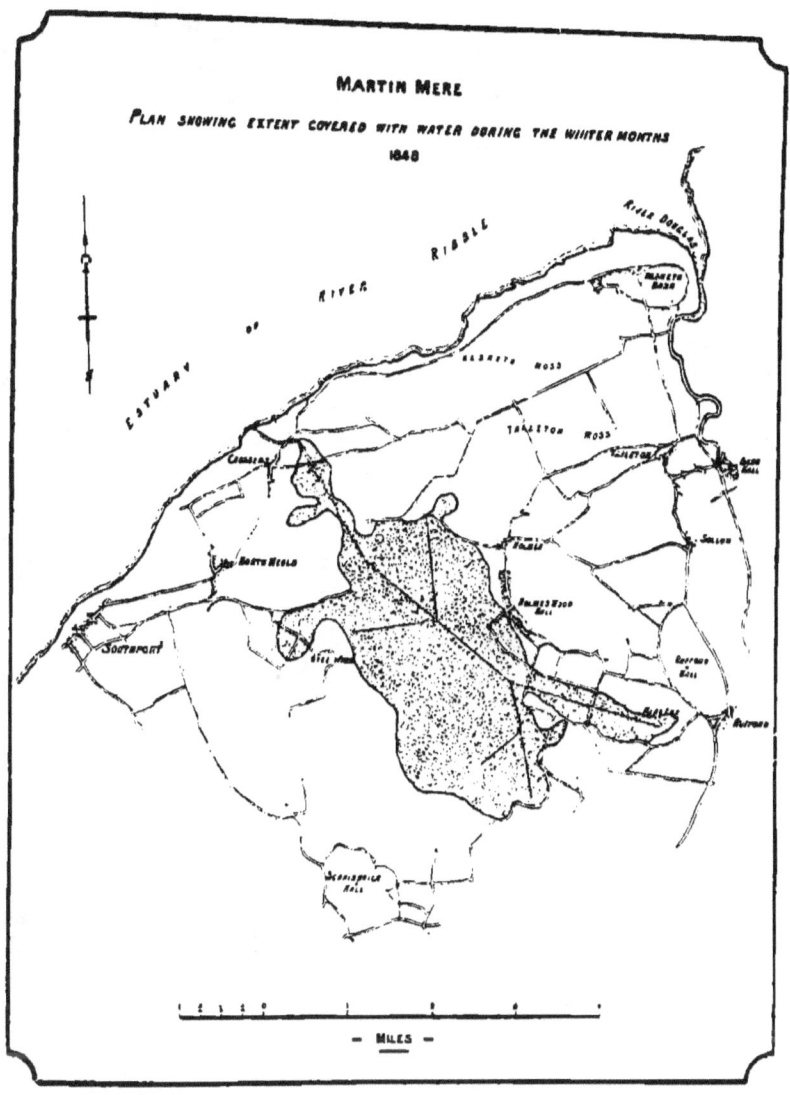

description, that "it is a bird common in these parts upon the sea-coasts, the shape of it is like that of other Crows, the head and wings being black and the body blue; its food for the most part are muscles [*sic*], and I have often with admiration observed these birds to peck [*sic*] up pebble-stones, and then to soar with them in the air to a considerable height, then to let fall the stones amongst the beds of shell-fishes which most commonly break some of them, they afterwards alight and feed upon their prey. These fowl are said to breed in the Isle of Man, but are not used as food."

ROOK.

Corvus frugilegus, Linnæus.

Local Names—*Craw, Crow.*

Resident, abundant and increasing, both on arable and grazing lands, and only falling off in numbers where suitable trees for nesting begin to get scarce. The neighbourhood of a dense population is no hindrance to the formation of a Rookery if any heavy timber be about, and in the first number of the *Manchester Guardian*, published on May 5, 1821, an account is given of the nesting of two pairs in that year in a small garden belonging to Mrs. Halls, at the top of King Street, one of the busiest parts of the town. The same place was tenanted in the following year, but Mr. Blackwall says (*Ann. Mag. Nat. Hist.*, 1847) that the birds finally left owing to Jackdaws, which had commenced

---

Mr. Ralph Johnson was Vicar of Brignal, near Greta Bridge, Yorkshire; he was the correspondent, friend, and assistant of Ray, and died May 7, 1695, aged 66 years.

building in the steeples of St. Ann's and St. Mary's, two churches in the vicinity, pilfering the sticks as fast as they brought them to their nests. In winter the Rookeries (which are almost always in trees which shed their leaves in winter) are deserted, and the birds may be seen night and morning flying to and from the woods of the district which are in the warmest and most sheltered situations. A spell of mild weather, however, always induces Rooks to visit the nests, and they may often be seen at such a time in December and January, repairing them, and bowing to each other on the branches in mistaken congratulations on the return of spring. The Rook is an early breeder, and lays its five eggs or so from the middle of March to the middle of April, according to the weather.

RAVEN.

Corvus corax, Linnæus.

Resident, and still breeding on some of the wilder hills in the north of the county. It is seldom seen on the lower ground except in severe frost, and, at such times, is as seldom permitted to return to its mountains. It used to breed in Wyresdale, and has done so within living memory at Thievely Scouts in Cliviger; sometimes yet being seen on Pendle Hill; while a few years ago a pair appeared in Bowland, but were hunted out of the neighbourhood. This or another pair bred at Bowland in 1886, but the nest was robbed on 14th March.—Ed. There are few of the hill-districts without some rocky crag which takes its name from this bird. Scattered examples of this species have been killed in winter

CLAP. OR CYMBAL-NETS.

almost everywhere, and it used to be not uncommonly seen on the sea-shore, but it is now exceedingly rare, and its extinction is probably only a question of a few years.

FAMILY ALAUDIDÆ.—GENUS ALAUDA.

SKY-LARK.

ALAUDA ARVENSIS, Linnæus.

Resident, and one of our commonest birds. It is universally distributed, and does not seem generally to vary much in numbers; but near Clitheroe it is not as common as it used to be, and Mr. R. Standen says the same of his district about Goosnargh. Individuals (whether the regular residents or not cannot be said) may be seen everywhere in winter, but a partial migration takes place in spring and autumn, and immense flocks collect on the sea-coast, very much added to by arrivals from north and south. Mr. R. J. Howard says that Sky-Larks and Starlings are the only birds which strike the lantern of Lytham lighthouse. Thousands are sent to the markets, caught on the mosses and sand-hills, mostly by means of what are called *pantles*: these being loops of horsehair attached at short intervals to a line fifty or sixty yards in length. William Blundell of Crosby, who was born in 1620, and whose jottings have lately been edited by the Rev. T. E. Gibson in "A Cavalier's Note-book," appears to have been a great bird-catcher, and besides relating modes of trapping wild-ducks, stares, hawks, and pheasants, says "a great help to the cymbal-nets* for bringing in of

* The cymbal-net or clap-net is still used very extensively on the stubbles and grass lands of Lancashire, mostly near the coast,

larks about your net is a gigg of feathers standing a distance off, which twirleth swiftly round on the least breath of wind. When the sun doth not shine, a foxtail pulled up within the compass of your net will make the larks strike at it as if it were a weasel." The Sky-Lark breeds commonly on the sand-hills, and on cultivated grounds it usually places its nest among the longer tufts of grass in the pastures, taking considerable pains at concealment. In summer it is the earliest morning songster (the Blackbird being the next), and Blackwall states that it sings from the 5th of February to the 16th of July; but I do not generally hear it very early in the morning before the warm days which often occur in April. Albinos are sometimes seen, though not

and the working of them is called locally "simmin" or "simblin." The *modus operandi* will be easily understood by reference to the illustrations. Two live Larks, called *brace-birds*, are fixed, one near each net and just outside, to one end of a lever which works on a peg, called the *brace-peg*, and from the other end a cord runs to the fowler's hand. A third cord is attached to the *mill*, which is a small stand, with a piece of looking-glass and a red rag fastened to it, and which revolves when the cord is pulled. The nets, the mesh of which is ten to the foot, are set on a bright sunny day, with the wind blowing directly into the face of the fowler. He sits on his empty box, and as soon as a Lark approaches, begins to whistle; its attention being attracted, he makes the brace-birds flutter, and twirls the mill, and when the wild bird has come sufficiently near, whether impelled by curiosity or what, the nets are rapidly pulled over, and the prize secured. In favourable weather, an average of eight or ten dozen a day, of which about eighty per cent. will be alive, can be taken; September and October being the only months in which birds "strike," as it is called, well. They are sold for from 1s. 3d. to 1s. 6d. per dozen, but cock-birds kept till after Christmas will fetch as much as 6s. per dozen, the destiny of these being the cage. Early in the season it is noticed that the Larks are smaller altogether than later (cf. Gray, "Birds of West of Scotland," 1871, pp. 122-23), but the fowlers profess to be always able to distinguish the cocks by their larger size.

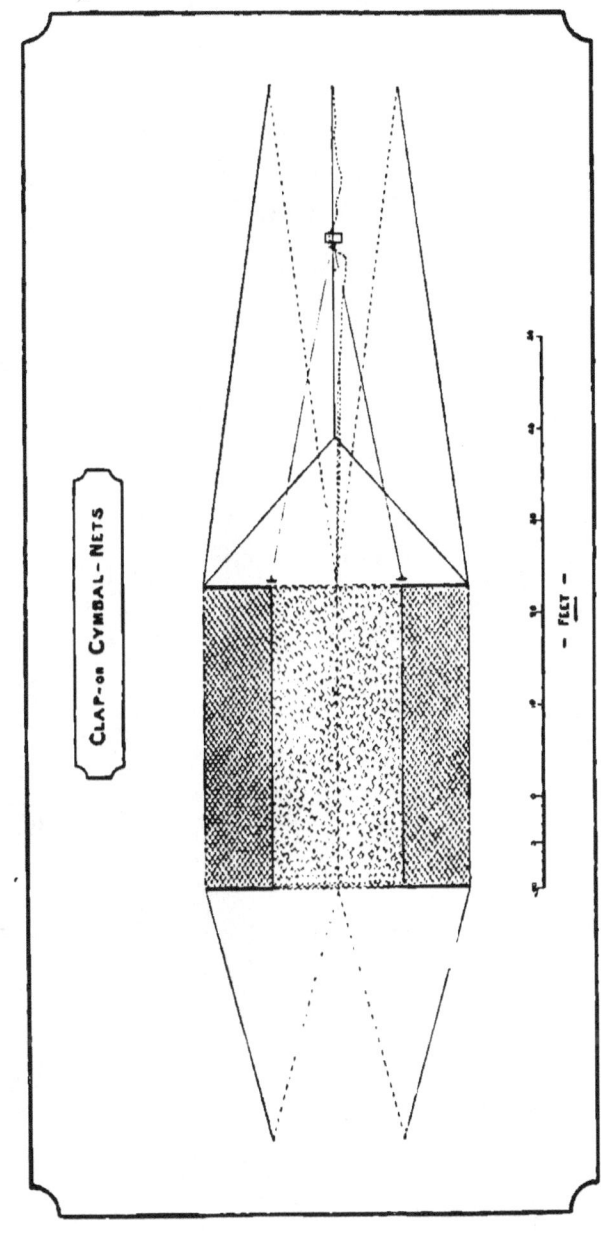

frequently, considering the immensity of the flocks (Mr. C. S. Gregson in the *Zoologist* for 1873 states that he has gone over in one house in one day seventy score dozens without seeing an abnormal feather), but melanic forms appear to occur oftener than with most other species. The Sky-Lark breeds the end of April or beginning of May, and lays four eggs in a nest of dry grass, this seldom being above the lower levels.

Mr. R. J. Howard says that a bird-catcher named Cookson found Sky-Larks more plentiful on Tarleton Moss in the winter of 1890-91 than he ever knew them before; in one fortnight he took 95 dozen.—Ed.

WOOD-LARK.

ALAUDA ARBOREA, Linnæus.

Once a common resident in many parts of Lancashire, the Wood-Lark is now almost extinct within the county boundaries, and is very rare at any season of the year. This appears to be the case generally throughout the north of England, and probably the incessant pursuit of this species by the professional bird-catcher has a good deal to do with its increasing scarcity. One or two score years ago it is recorded as having been plentiful near Liverpool and Prescot, in Wyresdale, at Samlesbury on the Ribble, and in the Winster valley: but no evidence as to any nest taken since that time has come under my notice, and hardly has its beautiful song been heard. Blackwall ("Researches in Zoology," p. 53) writes of it as singing in the neighbourhood of Manchester from the 20th of March to the 23rd of October: a mean of eleven years' observations, from 1818 to 1828, being taken.

Mr. R. J. Howard writes:—"On December 13th 1890, Ward saw a Wood-Lark in Livesey, feeding on spent hops with a flock of Greenfinches: he several times got within 20 or 30 yards of the bird, which at first attracted his attention by its note."—Ed.

GENUS OTOCORYS.

SHORE-LARK.

OTOCORYS ALPESTRIS (Linnæus).

An exceedingly rare winter visitor. Mr. John Hardy informs me that two instances of this bird having been shot on the Lancashire side of the Mersey have come under his notice: one in the winter of 1857 at Trafford, and another in 1873 near Didsbury; and Mr. C. S. Gregson, writing in September, 1882, says he recently got two alive and six dead caught in *gins* amongst Sky-Larks, this being on the coast near Formby. [The Rev. H. A. Macpherson records three examples, shot on Walney Island, October 29th 1890, by Mr. Arthur Bolton.—Ed.

ORDER PICARIÆ.

FAMILY CYPSELIDÆ.—GENUS CYPSELUS.

SWIFT.

CYPSELUS APUS (Linnæus).

LOCAL NAMES — *Longwing, Deviling, Devil-screamer, Devil-squeaker, Devil-skirler, Black Martin.*

A common summer visitor, appearing early in May, and leaving throughout August. The first departures take place in the beginning of the latter month, and birds are seldom seen after it is out, though Blackwall (*Mem. Manchester Lit. and Phil. Soc.*, 1824) mentions an instance in 1815, when he saw a Swift on the 20th of October. It is generally distributed in all localities, mostly in small numbers only, and those decreasing; but this I believe to be simply from want of suitable nesting-holes. It is much attached to ancient breeding-places, and in Clitheroe, where the species is very numerous, the same holes have been tenanted for very many years; some of them being in the roofs of mere cottages, and less than eighteen feet from the ground. In entering a new locality, however, only the loftiest buildings are chosen, and House-Sparrow and Starling are turned out without ceremony, and their nests appropriated; these being covered, after the occupancy of the Swifts has continued a while, by a sort of viscous matter which I presume exudes from their mouths. Mr. R. Standen says that near Longridge

Swifts breed in the stone quarries. They never build a nest for themselves, and unless one is seized ready made, the eggs are laid on the bare stones, the first week in June being the usual time for deposition, and two being the invariable number. Mr. T. Altham once found three eggs lying together, but as two were much incubated and the third was quite fresh, this last had probably been laid by another bird. Before their departure in autumn Swifts collect in considerable flocks, careering about in the evening, and filling the air with their screams; in the *Mag. of Nat. Hist.* for 1839 Dr. Skaife says that near Blackburn on July 25, 1838, he saw one of these flocks containing as many as several hundred individuals. It may be here remarked that the Swift is chronicled as arriving in 1774 at Blackburn on April 28 (White's "Nat. Hist. of Selborne," ed. Jesse, 1851, pp. 197, 271), and the observer was no doubt the Rev. John White, who was vicar of Blackburn from 1772 to 1780, brother of the naturalist, and himself the author of a "Natural History of Gibraltar" ("Hist. Blackburn," Abram, 1877, p. 296).

WHITE-BELLIED SWIFT.

CYPSELUS MELBA (Linnæus).

A rare straggler, of which the two following occurrences are known:—one, recorded by Mr. S. Carter in the *Zoologist* for 1863, the bird being captured in St. Mary's Church, Hulme, Manchester, on October 17 of that year; the other, killed near Preston in August, 1879, and now in the possession of Mr. Peter Sefton of Baxenden.

FAMILY CAPRIMULGIDÆ.—GENUS CAPRIMULGUS.

NIGHTJAR.

CAPRIMULGUS EUROPÆUS, Linnæus.

LOCAL NAMES—*Night-hawk, Flying Toad.*

One of the latest arrivals among the summer migrants, coming in May after the Swifts, and going in August; not being seen after that month, except in rare instances. It was once common on all the South Lancashire mosses, but is now only found on the more secluded ones, and where the character of these is suitable, and clumps of timber dot the heathy spaces, it breeds regularly. In the more northern parts of the county it shows greater partiality for the higher ground, and there it oftener frequents the woods on the low fells, becoming numerous in some parts of Furness. If disturbed during the day it is very languid in its movements, and I have seen it on the sand-hills of the coast so drowsy as almost to allow of capture by the hand. It does not rouse itself fairly till after sunset, and in June only commences to "jar" towards half-past nine, continuing to do so till about midnight, and hovering about in its butterfly-like flight in pursuit of the night-flying moths. At such times, too, it is very confiding, and will flit about within a yard of the observer, seemingly quite regardless of his presence. It lays two eggs, which are placed side by side on the bare ground among the heather, and begins to sit about the 10th of June. Mr. N. Greening says (*Nat. Scrap Book*, pt. 9) that on the mosses near Warrington he has seen it feign a broken wing in order to entice intruders from the vicinity of its nest.

FAMILY PICIDÆ.—SUBFAMILY PICINÆ.
GENUS DENDROCOPUS.

GREAT SPOTTED WOODPECKER.

Dendrocopus major (Linnæus).

A resident species, occurring at times in most of the thickly-wooded districts, and breeding regularly in some few. Byerley states that it is seen at Ormskirk and Knowsley, and it has been killed in almost all the woods round Manchester, having bred regularly near Middleton, according to Dr. Kershaw, up to 1875. A good many years ago a brood of young was taken in Witton Park, near Blackburn, and these are still in Gen. Feilden's possession. Mr. W. Naylor informs me that it used to occur at Clerk Hill, Whalley, and at Huntroyde, and Mr. H. Miller saw a nest of young near Whalley, on June 22nd, 1879, which were allowed to fly. The hole was in a decayed tree, and was more oval than round. In Mr. Cross's woods at Redscar it has bred for many years, and Mr. R. J. Howard tells me that in 1883 the young were duly hatched from a nest, and that in April 1884 the birds were frequenting their last year's nesting-place. Mr. R. J. Howard writes: "On May 12th, 1885, I found newly-drilled holes in a decayed ash near the river Roddlesworth, and on the 25th of same month I saw both old birds feeding their young at the nest-hole. From this date they nested yearly in the same tree, but on the 17th November, 1888, the ash was blown down, and although the birds are often seen, and I have found holes in several trees in Witton Park, I possess no evidence that they have bred in the immediate neighbourhood since."—Ed. Mr. W. Fitzherbert Brockholes says it has been killed in the Claughton Woods, and Mr. Chamberlain Starkie writes me that they have had it at Ashton Hall, whilst a pair frequented a wood in Bowland the whole of the spring of 1884. At Grange, according to Mr. Anthony Mason, it is occasionally seen, but appears to be very rare in Furness on the whole.

Authority has laid it down that no properly authenticated occurrence of the Middle Spotted Woodpecker, *Dendrocopus medius* (L.), has yet been recorded in Britain, but unless it be granted that the young of *D. major* preserves the crimson colour on the crown until the following spring, it seems difficult to reconcile the following note by the late Mr. Thomas Garnett with any other supposition. He writes in the *Mag. of Nat. Hist.* for 1822, "A pair of birds had hatched their young in a hole in a decayed ash, about twenty feet from the ground; there were two young ones, which I secured, as well as one of the old ones, and they are all now in the possession of a friend of mine. The old one measured $9\frac{1}{2}$ inches long, and weighed $46\frac{1}{2}$ dwts. an hour after it was killed: the forehead is a dirty buff, and the whole crown of the head a bright crimson." Mr. Tom Garnett informs me that his grandfather told him this nest had been taken near Clitheroe. The "Middle Woodpecker, *Picus medius*," mentioned by Pennant ("British Zoology," 1776-77) as having been shot in Lancashire, was probably the young in autumn of the Great Spotted Woodpecker.

LESSER SPOTTED WOODPECKER.

Dendrocopus minor (Linnæus).

Resident, but, as far as my information goes, now of very rare occurrence except in one or two localities. The Rev. J. D. Banister notes it as having been found at Redscar, near Preston, some thirty years ago, and Colonel H. W. Feilden writes me that about the same time one was shot by his father at Feniscowles. In the Report of the Bury Nat. Hist. Soc. (1871) it is stated on the authority of Mr. R. Davenport that this species has been known to breed in Simpson Clough, and several examples have been killed in that neighbourhood. Mr. Hugh P. Hornby tells me that one was shot at Winwick about 1867, and a year and a half ago Mr. W. Fitzherbert Brockholes saw one in his woods at Claughton, where in 1883, on May 11th, a nest of five eggs, a few days sat-on, was taken by Mr. Arthur Breakell. The bird was seen flying out of the nest-hole, which was in the trunk of an old oak, made by a branch being cut off and the part then decaying, and was about a yard up, and a foot into the tree. Mr. Anthony Mason says that the Lesser Spotted Woodpecker is occasionally observed near Grange, and in May, 1883, Mr. Henry Kerr saw one in the woods going to the top of Hamps Fell. Altogether this species would appear to be the rarest of its kind within the limits here treated of.

GENUS GECINUS.

GREEN WOODPECKER.

GECINUS VIRIDIS (Linnæus).

Times have changed very much since Dr. Leigh wrote in 1700 ("Nat. Hist. of Lancashire, &c.") that "The Heyhough* is common enough," and the Green Woodpecker is now only a rarely occurring species on the whole. It may still be occasionally seen in the neighbourhood of Grange, and perhaps nests there; and near Preston, where Mr. J. B. Hodgkinson says it used to breed commonly, it does so pretty regularly yet. In one or other of the thickly-wooded districts odd specimens are shot almost every year, but this appears to be generally in the winter season. It would be greatly to the advantage of forestry if species like the present were strictly protected by the land-owner. In the Ribble valley, for instance, the timber, especially larch, is suffering very seriously from the ravages of coleopterous larvæ; an evil which will become greater every year, and one to which the Green Woodpecker would apply a very efficacious remedy.

* Willughby ("Ornithology," Ray, 1678) says "The Green Woodpecker is by some called a Heyhoe, which name is, I suppose, corrupted from Hewhole, as Turner saith it was called in English in his time, and Mr. Johnson now."

SUBFAMILY IŸNGINÆ.—GENUS IŸNX.

WRYNECK.

IŸNX TORQUILLA, Linnæus.

Once a common summer visitor, now almost extinct. Blackwall gives its dates of arrival and departure near Manchester between 1814 and 1828 as averaging respectively April 11 and September 9; but Mr. W. Pearson, writing of the valley between Underbarrow Scar and Cartmel Fell, says (*Papers, &c.*, 1863) that in 1838 it appeared April 28th, and in 1839 on May 1st, and these last are probably the more correct dates, especially if any argument is to be derived from its sobriquet of 'Cuckoo's mate.' Even so far back as 1849 Mr. Pearson noticed it decreasing in his neighbourhood, for, writing on October 6th of that year, he says (*op. cit.*) that he has not heard its note for two or three seasons; remarking also on the shyness of its habits, and that its local name is "lang-tongue." Mr. J. B. Hodgkinson, too, speaks of its abundance formerly near Witherslack, and says that when he was a boy it frequently bred at Frenchwood, near Preston. Mr. C. S. Gregson informed Mr. A. G. More (*Ibis*, 1865) that he had once found its nest in Lancashire, and the Rev. J. D. Banister notes its occurrence in Wyresdale and at Stalmine. In Winmarleigh, in the second week of June, 1883, Mr. Arthur Breakell found a nest with seven fresh eggs, and watched the old bird for some time, as she kept flying from tree to tree, turning her head almost round, and all the time making a peculiar noise between croaking and hissing. The nest, of decayed bark and wood, was in the trunk of an old wild-apple-tree. This goes to show that the Wryneck might again become abundant as a breeder, but in most localities it has for many years been represented only by an odd specimen here and there, shot or picked up dead.

FAMILY ALCEDINIDÆ.—GENUS ALCEDO.

KINGFISHER.

ALCEDO ISPIDA, Linnæus.

A resident species, and—in spite of the incessant pursuit which its beautiful plumage entails upon it—still numerous on all the more secluded streams. A few years ago, owing to the prevailing fashion, this pursuit became so severe, at least in my own district of Clitheroe, as almost to endanger its existence; but a rapid recovery has taken place, and in 1883 there were nine pairs of birds within a very small area. The large number of eggs (six or seven) which the Kingfisher lays, is in its favour, for the successful bringing up of a single nest forms a considerable nucleus for next year's propagation. In the south of the county it is oftener seen in winter than at any other time, and the increase of population has driven it from many formerly favourite localities near Liverpool and Manchester. Mr. Clayton Chorlton writes in the *Manchester City News* of 1882, that four or five years before, the straightening and levelling of the banks of a small brook destroyed the last refuge of the Kingfisher in Withington, and it is now seldom seen on, at least, the lower reaches of the Irwell and its tributaries. On the higher ones, however, a few pairs still remain, and going north it is found breeding in more or less numbers on almost all the rivers and brooks. The Ribble and Hodder, the Wyre and Cocker, the Winster and Duddon, are all frequented by this lovely bird, and though a partial migration to the coast always takes place in winter, odd specimens may be seen throughout the year, except in the very severest weather. The Rev. J. D. Banister writes in his journal, of the Pilling district, that the Kingfisher " seldom nests with us. In September or early October it invariably appears on our coast, and may be found on the pools on our marshes or in the inland watercourses immediately adjoining the sea. From September to March I could any day find several in a short time, while in the remaining part of the year it is a very great rarity to find one within several miles." Mr. W. A. Durnford also notes it as more numerous on the shores of Furness at the end of the breeding-season, and writes (*Zoologist*, 1876) that a few years before a specimen was killed by flying against the lighthouse on the south end of Walney. It invariably, I believe, digs a fresh hole for itself every year, and does not occupy an old one, unless driven from the first, and its eggs or young destroyed : the same bank will be bored year after year with fresh holes if no interference takes place. Immediately the eggs are laid, undigested fish bones are voided all round the sitting bird, and an increasing ring is made whilst the eggs are being hatched and the young are growing ; the stream of faecal matter which then flows to the entrance of the hole rendering it exceedingly offensive. It may always be taken for certain that if no cast-up bones are about the mouth of the hole no eggs have been laid. The Kingfisher is an early breeder ; beginning to dig late in March or early in April, and having the full complement of eggs deposited by the third week of that month. Each pair seems to appropriate to itself a certain length of stream, and it is seldom that more than two birds are seen within half a mile of each other.

FAMILY CORACIIDÆ.—GENUS CORACIAS.

ROLLER.

Coracias garrulus, Linnæus.

An occasional, but rare, visitor in summer. Pennant mentions one being shot at Dalton-in-Furness on May 26th, 1827, and in 1836, as I am informed by Mr. J. B. Hodgkinson, one was obtained at Walton-le-Dale. Byerley ("Fauna of Liverpool," 1856) says that a Roller was shot at Knotty Ash, but gives no date, and Mr. W. Naylor writes me that one is still preserved at Burnley, shot at Marsden near there, on August 25th, 1848. Mr. A. Wood, of Simpson Clough, has a specimen killed in June 1860, on Walney Island, and in May 1868, one, in poor condition, was shot at Blackpool, being now in the possession of Mr. R. Drummond.

FAMILY UPUPIDÆ.—GENUS UPUPA.

HOOPOE.

Upupa epops, Linnæus.

A rare visitor from spring to autumn. The following occurrences are noted: One shot at Longton, near Preston, September 23rd, 1841 (John Skaife, *Ann. Mag. Nat. Hist.*, November, 1841): One seen at St. Michael's-on-Wyre in summer of 1849 (Hugh P. Hornby): Four shot at Knowsley, now in the Derby Museum, Liverpool; one at Edge Hill; at Formby and elsewhere (Byerley, "Fauna of Liverpool," 1856): One shot at Knott End, about 1864, by R. Croft (H. Miller): One, a young male, killed near Ashton-under-Lyne in May, 1865 (C. W. Devis, *Zoologist*, 1865): One, a male, shot at Garston early in September, 1867, and one at Everton a few years previously (*Liverpool Naturalist's Journal*, October, 1867): One shot at Ringley Moss, July 3rd, 1869, a second being seen in the woods adjoining the Park, Pilkington, the same week (R. Davenport): One, a male, obtained near Cuerden Hall, Preston, August, 1875 (H. Shaw, *Field*, August 28, 1875): One on the Ribble at Redscar, some years ago (J. B. Hodgkinson).

[The Rev. H. A. Macpherson adds: One, Walney Island, in the spring of 1884.—Ed.]

FAMILY CUCULIDÆ.—GENUS CUCULUS.

CUCKOO.

Cuculus canorus, Linnæus.

This familiar summer visitor is stated by Blackwall ("Researches in Zoology") to arrive in the neighbourhood of Manchester on the 20th of April, and to leave the 27th of June, the mean temperature in the shade on the former date being 47°, and on the latter 59°, and the observations extending over the fifteen years from 1814 to 1828. From the comparison of a very large number of notices which have reached me, I should conclude that, on the whole, about a week later, or the 27th of April, is nearer the average; but the dates of course vary with the character of the locality from which the records come. For instance, in the sheltered districts at the head of Morecambe Bay, the Cuckoo appears very early, and in 1883, Mr. H. Kerr informs me that near Grange it was heard regularly after the 10th of April, while from no other district have I any note of its arrival before the 28th. The earliest reliable date I can find is that recorded in the *Field* of April 27, 1861, in which year Mr. John Page says that it was seen near Manchester on the 2nd of April. The old birds are seldom observed after the first few days of July, and the young of the year the same time in September. The Cuckoo is universally distributed, and is equally common on the South Lancashire mosses, the sand-hills of the coast, the cultivated lands, and the bases of the hills in the moorland districts. It lays its eggs from the 1st of May to the middle of June, and the nest of the Meadow-Pipit is, undoubtedly, oftenest chosen for the purpose, the Sky-Lark, Hedge-Sparrow, Yellow-hammer, Whinchat, Pied and Yellow Wagtails, and Sedge-Warbler being, in their order, the next most favoured recipients of its attentions. The call of the male Cuckoo is frequently heard in the night, sometimes even into the small hours.

ORDER STRIGES.

FAMILY STRIGIDÆ.—GENUS STRIX.

BARN- or WHITE OWL.

Strix flammea, Linnæus.

Local Names—*Hullet, Howlet.*

The Barn-Owl has been now for a long time recognized by the farmer as a valuable ally against the mice which infest his fields and barns, and the care with which it is preserved against intruders when nesting has almost become a superstition. The gamekeeper, however, holds different opinions, and although this species is not harassed like the true birds of prey, its numbers are very much reduced by him; and owing to greater strictness in game-preserving, there is probably not more than one nest now, where twenty years ago a half-dozen might have been found. This Owl is resident throughout the year, and is very generally distributed, being found in limited numbers in all the country districts; breeding there most commonly in isolated barns. The pellets of undigested fur and bones which it emits, form after a while an increasing heap about the nest, and in some long-frequented recesses these pellets would make a good barrow-load. The eggs are from five to eight in number, and are laid, but not on consecutive days, about the end of April or beginning of May. Mr. J. Murton (*Zool.*, 1870) mentions an instance in which he found eleven eggs, but these might, perhaps, be the produce of more than one bird. Incubation begins when the first egg is deposited, no doubt resulting in part from the fact that the old bird remains upon the nest all day. Mr. John Weld (who says that a few years since the White Owl bred in every barn in Leagram, but has now become very scarce) thinks that the last eggs are hatched by the first young. The call-note is a shrill whistle, for this Owl never hoots. I am able to confirm in part Waterton's observation that it will sometimes catch fish, for I once saw a pair flying backwards and forwards over the Ribble, and repeatedly dropping on to the water with a great splash, and then rising again, but I could not be sure whether they were fishing or only washing themselves.

FAMILY ASIONIDÆ.—GENUS ASIO.

LONG-EARED OWL.

Asio otus (Linnæus).

Local Name—*Horned Owl.*

The Long-eared Owl is resident, but in the breeding-season very local: it is probable that a partial migration takes place in autumn, for at that season and in winter specimens have been seen and shot almost everywhere. Large and retired woods are its favourite habitat, and in those at Knowsley its nest has been several times discovered. Mr. J. J. Hornby has an egg taken there in 1880 on the 20th of March, from a nest in a Scotch fir which he concluded was an old Wood-

Pigeon's, and he thinks there are always a few pairs breeding annually in the Park, as he regularly sees birds in the spring among the fir-plantations. Lord Lilford writes me that it breeds in the fir-woods at Rufford, and, according to Mr. R. Standen, it does so also at Claughton. Mr. T. Jackson says that a nest was taken at Winmarleigh in 1880, and that every year four or five specimens pass through his hands. It is rare in the Ribble and Hodder valleys, and no instance of its breeding has come under my notice within the last ten years (about which time ago Mr. W. Peterkin tells me a nest of young was found at Whitewell) until the present year, 1884, when on May 16th, the old birds were shot, and four young, a fortnight old, taken, on Longridge Fell from a Magpie's nest of a former season. A little later also, in the same locality, four eggs were taken, and the old bird shot, from another nest. Mr. W. A. Durnford says that a few pairs nest annually near Barrow, and that birds are often shot in winter, they being unusually plentiful in that of 1875 (*Zool.*, 1876). The Long-eared Owl feeds very largely on birds, and in the stomach of one which Mr. W. Naylor had brought to him on March 30th, 1880, was a Thrush of some kind, almost whole. Mr. H. Ecroyd Smith (*Proc. Histor. Soc. of Lanc. and Chesh.*, 1865–66) also says that in a number of pellets he examined near a tree in which was a nest of young, in May, 1865, a large proportion contained the skulls of young House-Sparrows.

SHORT-EARED OWL.

Asio accipitrinus (Pallas).

Local Names—*Moor-Owl, Brown Owl, Fern-Owl, Grey Hullet.*

The Short-eared Owl is by far best known as a winter visitor, but it still remains to breed in one or two localities, and would do so, without doubt, in many more, if only permitted by the gamekeepers. It arrives about the last week of October, frequenting at first the sand-hills of the coast, and thence straggling to the more inland wastes, mosses, and moors, on which, in all parts of the county, specimens have been shot. It is especially common on the island of Walney, but does not seem to remain to breed [A pair bred there in 1884, as I learned on the spot in 1885.—Ed., and disappears in February or March. Mr. W. A. Durnford, who has had great opportunities for observation there, says (*Zool.*, 1877) that in the winter of 1876 it was unusually plentiful, and that during November numbers were killed almost daily. In this year it was also noticed as being more numerous than usual in the Southport district, but the winter following Mr. Durnford found it just as scarce at Walney as before it had been the opposite. He records in the *Field* of June 19, 1880, an instance of its breeding on the borders of Lancashire and Westmorland, there being six young birds in the nest, all of different sizes, and young were reared, too, in the same place in 1884. It has bred in Bleasdale, and a nest, containing six fresh eggs, was taken on the lower slopes of Pendle Hill in May 1877, by Mr. T.

Jones, but the mosses in the vicinity of the Ribble estuary are the stronghold of the Short-eared Owl in the breeding-season. Here, among the heather, it nests annually, and in 1883, Mr. John Sumner tells me he saw in the middle of May a nest with six eggs, and about the first week in June one with six young. Mr. R. J. Howard writes me that on the Scarisbrick estate it breeds occasionally, and that in winter as many as twenty may be flushed from a single field, generally one of rough grass or turnips. Its services in clearing off the field-mice are gratefully recognized by the farmers, but the fenmen are much troubled by its practice of carrying off the Snipes and Larks which are caught in their pantles and other snares.

GENUS SYRNIUM.

TAWNY OWL.

SYRNIUM ALUCO (Linnæus).

LOCAL NAMES—*Wood-Owl, Brown Hullet.*

Resident, but decreasing, and now confined to a few thickly-wooded districts, the neighbourhood of a river being especially preferred. From the southern division of the county I have no records whatever, but up the Ribble, Hodder, and Calder, and in Wyresdale a few pairs breed here and there every year, and in Furness also, according to Mr. W. A. Durnford, it is resident, but not numerous. Writing in the *Mag. of Nat. Hist.*, 1838, Dr. Skaife says that it has become rather scarce, and that he learns from the gamekeepers, then their bitterest persecutors, that within a few years they were very numerous in the woods towards the Ribble. If every bird that takes game occasionally is to be destroyed, then is the Tawny Owl doomed to extinction, for there is no doubt it sometimes sins in this way; but as it is when their young are being brought up that birds of prey commit the greatest havoc among game, and as this species hatches its eggs very early in the season, it is much less destructive to the preserves than it otherwise would be. It does not build a nest for itself, but oftenest chooses a Carrion-Crow's of a former year, merely adding a few of its own feathers as a lining. The eggs are four in number, and are laid the last week in March or beginning of April; incubation commencing, as with the other Owls, immediately on the deposition of the first egg.

GENUS NYCTALA.

TENGMALM'S OWL.

NYCTALA TENGMALMI (J. F. Gmelin).

With reference to the example mentioned by Professor Newton ("Yarrell's British Birds," 4th ed., vol. i. p. 155) as having been seen by him at Nottingham, Mr. William Felkin, now of Auckland, New Zealand, writes me under date April 18, 1884, enclosing an extract from the account he kept of all the birds in his collection, and which he still has. It runs thus: "Tengmalm's Owl. The specimen in my collection of this bird was obtained by me from a weaver of my acquaintance at Preston, Lancashire. I called on him when passing through, on my way to the Lakes, in the summer of 1851. He brought it in fresh shot near Preston, I think close to Penwortham." The bird is now in the Nottingham Town Museum.

GENUS SCOPS.

SCOPS-OWL.

Scops giu (Scopoli).

Mr. John Plant writes me that a specimen, now in the museum of Peel Park, Salford, and acquired by purchase in the year 1850, was stated to have been trapped in Boggart Clough some years previously.

GENUS ATHENE.

LITTLE OWL.

Athene noctua (Scopoli).

The only recorded occurrence of the Little Owl in Lancashire is that by Mr. Thomas Williams, Bath Lodge, Ormskirk, who, asserting his acquaintance with the species, wrote as follows in the *Naturalist's Scrap Book*, 1863, pt. 5: "A few years ago I saw a fine specimen of this beautiful bird in the Bath Wood, Ormskirk. It was perched on a dead stump near the ground, and seemed quite regardless of my approach, which was accidental. It was broad daylight in the summer time, and the little creature seemed spent or stupefied, whether wearied with long flight or in that drowsy state so common to Owls in the day-time I cannot say. I could have taken it with the hand, and am certain of its identity."

ORDER ACCIPITRES.

FAMILY FALCONIDÆ.—GENUS CIRCUS.

MARSH-HARRIER.

CIRCUS ÆRUGINOSUS (Linnæus).

The Marsh-Harrier is now only an accidental visitor, and is very rarely seen. About twenty-five years ago Mr. R. J. Howard says a hen-bird was shot from the nest on Martin Mere, near Southport, by a man named John Cookson, and the Rev. J. D. Banister states in his MSS. notes that it bred formerly at Pilling. Byerley ("Fauna of Liverpool," 1856) writes of it as occurring "in the rabbit-warrens about Crosby and Formby (Mather)," and Mr. John Watson asserts that not long ago it bred on one of the low-lying mosses to the north of Morecambe Bay, close to the border line of Lancashire and Westmorland. An example is still preserved at Leagram Hall, shot by the late Mr. Weld many years ago, on the suggestive date of August 12th.

HEN-HARRIER.

CIRCUS CYANEUS (Linnæus).

The Hen-Harrier (commonly known as the Blue Hawk from the plumage of the male, and as the Ringtail from that of the female) still breeds occasionally in Higher Wyresdale, and Mr. R. Standen vouches for a nest of four eggs having been found there about the year 1876, in the month of June. The shepherds see birds about pretty regularly, and Mr. J. B. Hodgkinson tells me that he has often had specimens sent him in breeding-time, while the same locality is mentioned also by the Rev. J. D. Banister as frequented by this Harrier. Elsewhere I have no record of its breeding, but Byerley says it has been shot in many places near Liverpool, and Mr. J. F. Brockholes writes (*Proc. Liverpool Lit. and Phil. Soc.*, 1859–61) that a kind of Harrier may be often met with, which the keepers call the Blue Hawk, and which he has never been able to identify, but which I take to be this species. The Blue Hawk is known, too, by the fowlers of Martin Mere; and an immature male of the Hen-Harrier which was accompanied by a female, was, as I am informed by Mr. R. J. Howard, shot about twenty years ago at Rufford by Henry Caunce, and is still in his possession. Dr. Skaife (*Mag. Nat. Hist.*, 1838) considered it very rare then, but mentions his having a specimen of the male, which was shot near Lancaster. Mr. H. Kerr says that it has been shot on passage in the Rossendale district, and in autumn Mr. Standen is accustomed occasionally to see it hawking over the stubbles near Goosnargh.

MONTAGU'S HARRIER.

CIRCUS CINERACEUS (Montagu).

Mr. W. A. Durnford writes (*Zool.*, 1876) that one, shot on Walney in the autumn of 1874, is in his possession, and this is the only occurrence I find recorded.

Mr. R. J. Howard writes:—" 15th June, 1889.—

Mr. J. Gregson, gunsmith, brought a female Montagu's Harrier, shot by Mr. Eli Heyworth's keeper on Whitendale Moor, on 10th inst.; hatching spot bare. The keeper said that the bird was seen to dash at the head of a sheep; this was done, doubtless, to drive it away from the nest, for I feel sure the bird was breeding. Mr. Heyworth gave me the bird, of which I have skin and sternum."

GENUS BUTEO.

BUZZARD.

BUTEO VULGARIS, Leach.

As a breeding species the Common Buzzard has been exterminated in every part of Lancashire except those few peaks in the north which lie within the county, and which form a portion of the Lake mountains. There a few pairs still remain, but Dr. C. A. Parker of Gosforth, who has paid great attention to raptorial birds, considers them decreasing; and indeed, considering the persecution they suffer, it is a wonder that any are now in existence at all. In the Forest of Bowland, and generally on that wild range of hills which lies between the valleys of the Ribble, Wyre and Lune, the Common Buzzard is shot or trapped almost every year. The Rev. J. D. Banister, writing between 1840 and 1850, says that it "breeds in Wyresdale," and Mr. John Weld is confident that a nest which was taken in Buckbanks in Leagram a few years ago was of this species. In a paper read before the Kendal Nat. Hist. Soc. on December 8, 1839, Mr. W. Pearson says that a few years before it bred in the larch plantations at Lamb How, but that it had been exterminated in the interests of sport: he further remarks on one of a singularly piebald appearance that haunted the woods and wastes near Gill Head in Cartmel Fell at that time, and also that the stomach of one he once shot was filled with earth-worms. During the winter months, and especially in hard weather, the Buzzard often descends to the vicinity of the coast, and at this season it has several times been killed on Walney, and the sand-hills between Liverpool and Southport. It has also straggled occasionally further inland, and on these occasions has been found to have been feeding principally on mice and rabbits.

["From November 1886 to March 1887 a Buzzard frequented Billinge Hill, feeding chiefly on some slaughter-house garbage spread as manure on a field near, and roosting at night in the trees on the hill. I saw the bird often: it was an interesting sight to watch him wheeling about high over the factory chimneys. The head-keeper, contrary to express orders, shot at the Buzzard and broke a leg on 8th January 1887. For some weeks afterwards I saw the leg hanging as the bird flew, but it gradually recovered, and when I last saw the bird on 24th March it seemed all right."—R. J. H.]

GENUS ARCHIBUTEO.

ROUGH-LEGGED BUZZARD.

ARCHIBUTEO LAGOPUS (J. F. Gmelin).

An autumn migrant; usually of rare occurrence, but some years appearing in considerable numbers. In 1880, for instance, there were five specimens shot; more than I had found recorded for many years previous altogether. They were as follows; all, as far as I could ascertain, being immature birds of the first year:—A male, shot October 11th, on Haslingden Moor, by Mr. John Hoyle 'Mr. Hoyle told Mr. Howard that this bird was being chased by a Merlin.—Ed.', (Howard, Zool., 1880, p. 488): One, shot October 14th (S. C. de Trafford, *Field*, October 30th, 1880): A female, shot on Waddington Fell, near Clitheroe, October 30th (this was just over the Yorkshire border), stated in the local papers to have been a Snowy Owl: A female, on November 23rd, which had been eating rabbits in traps (John Wrigley, Formby, *Field*, January 15th, 1881): One, shot near Blackpool in December (R. Drummond): [One near Bolton, C. E. Stott, *Zool.*, 1889, p. 77.—Ed.]

GENUS AQUILA.

SPOTTED EAGLE.

AQUILA CLANGA, Pallas.

Mr. W. A. Durnford ("Birds of Walney," 1883) says that in 1875 he examined one of these rare and handsome birds, which had been picked up dead by some fishermen on the west shore of Walney Island, but was not able to ascertain any further particulars in connection with its occurrence.

GOLDEN EAGLE.

AQUILA CHRYSAËTUS (Linnæus).

One shot near Millwood, close to Furness Abbey, in 1815, W. B. K. (Durnford, "Birds of Walney," 1883).

GENUS HALIAËTUS.

WHITE-TAILED EAGLE.

HALIAËTUS ALBICILLA (Linnæus).

An accidental visitor, of rare occurrence. At Leagram Hall is preserved a specimen, in immature plumage, which was killed when roosting in Hodder Hole Wood by some poachers in November, 1840, and Mr. John Weld tells me that Eagles, probably of this species, used occasionally to be seen at intervals of a few years in that neighbourhood, or on the adjoining high moors, in hard winters after very stormy weather. A Sea-Eagle was taken alive in the year 1838 near Broughton-in-Furness [The Rev. H. A. Macpherson states that this bird was really captured on the top of Blackcombe, Cumberland.—Ed.], and was kept in confinement at Broadgate, the residence of the Lewthwaite family, until 1846 or 1847, when it died through an injury to its wing, inflicted during recapture after an attempt to escape from the garden where it was confined. The feet were preserved, and Mr. William Lewthwaite has been kind enough to forward them to me for the purposes of identification. Mr. C. S. Gregson says (*Nat. Scrap Book*, 1864, pt. 16) that he has examined an immature Sea-Eagle, shot on Blundell sands some time before, and Mr. Hugh P. Hornby writes me that he is convinced it was a bird of this species which he saw at St. Michael's-on-Wyre in the last week of October, 1875, and the whiteness of its tail led him to believe it was adult: he says that close to the spot whence it rose was a dead rabbit with eyes and entrails cleaned out.

GENUS ASTUR.

GOS-HAWK.

Astur palumbarius (Linnæus).

Of very rare occurrence. One was shot near Colne in 1863, according to Mr. Henry Whalley, and Dr. Skaife writes as follows in the *Mag. of Nat. Hist.*, 1838, "Very rare, though shot or caught occasionally in the Forest of Bowland. A relative of my own has a beautiful pair, male and female, caught in a trap there a few years since."

GENUS ACCIPITER.

SPARROW-HAWK.

Accipiter nisus (Linnæus).

The Sparrow-Hawk is a resident species, and is found in all wooded districts, breeding more or less commonly in proportion to the energy shown by the gamekeepers in its extermination. It may be seen in the hardest winters, though then much scarcer, and it is very probable that such birds are migrants from more northern regions. More wary than the Kestrel, it is more persecuted, and of all my informants only Mr. C. E. Reade of Urmston considers that in his neighbourhood the Sparrow-Hawk is the commoner of the two. It almost invariably builds a nest for itself, and returns to the same every year; should this be destroyed in winter, a new one will be constructed, but if this be robbed, then the nearest deserted Magpie's or Carrion-Crow's will be occupied, and a few fresh strips of bark just added as a lining. The nests are never very high up, and indeed are so flat that, if much exposed to the wind, their contents would be seriously endangered. Like the rest of the Hawks, this species does not lay an egg every day, nor does it begin to sit on the first egg, though it will do so after two or three have been laid. Mr. T. Altham has taken eggs as early as the 16th of April, but the beginning of May is a more usual time, the dates varying with the seasons. He has seen eight eggs in a nest once, and six is not an uncommon number, but the average is not more than five. The Sparrow-Hawk is sometimes very savage when her home is being disturbed, and I have seen the hen sit screaming on the nest-edge whilst the climber was ascending the tree, and fiercely strike at his head or hand as soon as he came within reach. It feeds mostly on small birds and the larger insects, and Mr. B. N. Peach states that once, when lying on the grass, a male bird alighted within a few yards of him, and shuffling along in the usual ungainly fashion, caught the crane-flies, which swarmed there, in one foot, and transferred them at once to its mouth.

GENUS MILVUS.

KITE.

MILVUS ICTINUS, Savigny.

The Kite has long been exceedingly scarce, and now only occurs very infrequently. At one time it must have been well known, for William Blundell of Crosby, as quoted in "A Cavalier's Note-Book" (ed. Gibson), writes of it in the seventeenth century very familiarly. He says that if you "take a Kite and a Carrion-Crow, and tie them down in the stubble with sufficient liberty, they will fight and cry in a strange manner: upon which there will come immediately great flocks of Crows from all parts, which striking freely at the Kite will many of them be taken in the lime twig which must be placed round in the stubble for that reason. Remember that you tie up the foot of your Kite to make the battle more equal. You may easily take a Kite with a Pigeon and lime." Probably the last instance of its breeding in Lancashire is that recorded by Mr. W. Pearson (*Papers, &c.*, 1863) who, writing in 1839, says that he "well remembers the Kite in his youth, his forked tail and his long crooked wings," but that "it has not been seen in Crosthwaite for thirty years:" he continues, "my neighbour, Isaac Walker, informs me that when he was a youth, fifty years ago, and lived in Sawrey, a pair or two of gledes built their nests among a number of tall trees, on the west side of Windermere lake, near the Ferry Inn. These birds were most of them destroyed by some idle fellows in the neighbourhood, who shot them on the roost during moonlight. He once took away a young one from a nest containing two: it became very tame, and would sit upon his hand; and although it had long and sharp talons, it took care never to hurt him with them. He permitted it to be at liberty, and it would sometimes stay away for a day or two, but always returned. At length, however, it entered the cottage of an old woman without leave, and the ill-natured crone killed it." Mr. W. A. Durnford's correspondent, W. B. K. ("Birds of Walney," 1883), says that the "Glead" used to breed in Low Furness; and according to the Rev. J. D. Banister and Mr. John

Weld, it at one time was occasionally to be seen in Wyresdale, and on the Bleasdale Fells, but is rarely met with now. In 1868, Mr. R. Standen saw a male bird which had been shot near Inglewhite in September of that year, and Mr. J. Clayton Chorlton writes me that about 1876 a specimen was shot in the neighbourhood of Manchester.*

GENUS PERNIS.

HONEY-BUZZARD.

PERNIS APIVORUS (Linnæus).

Byerley writes of the Honey-Buzzard: "a dozen at least from about the district of St. Helen's, Aintree race-ground, and elsewhere (Mather), Rainford, 1835." No other localities have been so favoured, and the remaining occurrences, which appear to have all been so far as is known in autumn, are very few. Mr. R. Davenport says one was killed in 1852 on Lostock flats by R. Shaw, and in September of the same year, according to Mr. Henry Whalley, one was shot near Colne. About 1860 one was obtained near Burnley by a keeper in the employ of General Scarlett, and a correspondent of the *Field* writes (November 10th, 1866) that he shot a fine specimen while out Grouse-driving on Blackstone Edge on the 8th of October.

* In the *Field* of June 22nd, 1861, the late F. T. Buckland wrote that at a sale of the Macclesfield Museum on June 14th, a Swallow-tailed Kite, *said* to have been shot on the Mersey in June 1843, fetched £9 10s. [If by this the American *Elanoïdes furcatus* is meant, there is no evidence that the bird was obtained in a wild state, but it may have been brought over in a ship.—Ed.]

GENUS HIEROFALCO.

GREENLAND FALCON.

Hierofalco candicans (J. F. Gmelin).

In the Proceedings of the Historic Society of Lancashire and Cheshire, session 1865-66, is published a note by Mr. C. S. Gregson, who says that an old bird is in his possession, which flew on board a vessel coming into the port of Liverpool and was captured by one of the crew.

GENUS FALCO.

PEREGRINE FALCON.

Falco peregrinus, Tunstall.

Like the Common Buzzard, the present species still manages to carry on a precarious existence on some of the Lake mountains, and, in spite of persecution, probably two broods are hatched there each year. This number has not been exceeded for a long time, and the case on the Lancashire side of the border is no doubt identical with that on the Cumberland hills, where Dr. C. A. Parker says the Peregrine neither increases nor decreases. The only other locality as to which there is evidence of its nesting is that of the gorge of Cliviger, where is a rock still called Eagle Crag,* and which, like

---

* The name has been said to have been given the crag from its resemblance to an Eagle's beak, but this, I think, is fanciful. It is also known as Bill Knipe, probably derived in part from *Knipe*, which is Norse for a projecting rock.

many others, thus mutely bears witness of its former tenants. Dr. Whitaker ("Hist. of Whalley," ed. 1818, p. 345) mentions this, and as his family seat of Holme is close by, he no doubt wrote from personal experience. Speaking of Cliviger, he says, "The almost inaccessible rocks above resound with wild and various yells of Hawks, which inhabit these secure retreats, to the destruction of vast quantities of game, whose bones form little charnel-houses about their nests. Among these one pair, of far superior size and strength, popularly called Rock Eagles, but really the Peregrine Falcon, now become extremely scarce, have annually bred from time immemorial." I have examined a bird which was shot here in 1820 by the late Vicar of Whalley, the Rev. R. N. Whitaker, and Mr. W. Naylor tells me that some eggs, forming part of a very old collection, which passed some time ago into the possession of Colonel H. W. Feilden, and which I have seen, were also from this crag. The Peregrine has been shot as a straggler at all times of the year, but oftenest on migration in spring, in various parts of the county, and on the wild hills of the Forest of Bowland it is not at all an uncommon visitor at that season.

HOBBY.

Falco subbuteo, Linnæus.

Specimens of the Hobby have been procured in Lancashire both in winter [Mr. Howard has examined an adult male, in the collection of Mr. Drummond of Blackpool, shot by Lady Clifton's head gamekeeper late in November or early in December 1885: a very mild autumn.—Ed.] and summer, but it only occurs very

EAGLE CRAG OR BILL KNIPE, VALE OF CLAVIGER.

rarely. Mr. John Plant informs me that one was shot at Barton-on-Irwell between 1830 and 1840, and in the latter year it is stated in the Report of the Bury Nat. Hist. Soc. (1871) that one was got at Simister Lane, being now in the possession of Mr. Wright Johnson of Prestwich. The same Report records the capture of one at Bolton in 1870, and in a note supplied in 1871 to Mr. Dresser's "Birds of Europe," Lord Lilford writes: " A young Hobby was shot in South Lancashire some years ago by my father's gamekeeper, who told me that it followed his pointer dog, as he was Partridge-shooting, for a considerable distance, and kept stooping and striking the dog till he was quite disgusted, and came in to heel." Byerley says, " Specimen shot at Knowsley, in the Derby Museum. One at Crosby,—Mather," and Mr. C. S. Gregson notes (*Proc. Histor. Soc. Lanc. and Chesh.*, 1865–66) that in 1865 he was shown a male bird which had been shot on Cuerdley marsh. Mr. R. J. Howard informs me that he has examined a female at Shaw Hill, Chorley, which was shot in the middle of June, 1878, by Colonel Crosse's gamekeeper, in the act of carrying away a young Pheasant, and the Rev. J. D. Banister writes that the Hobby has occurred in Wyresdale forest, but does not give any particulars.

MERLIN.

Falco æsalon, Tunstall.

Local Names—*Little Hawk, Tweedler.*

The Merlin is found breeding in small numbers on all extensive tracts of moorland; occurring in winter most commonly near the sea-coast. It appears to be very local, and will return to the same patch of hill-side year after year, however much disturbed it may have been in previous seasons. Each pair appropriates a very considerable stretch of country, and I have never known nests at anything but a long distance from each other. Although its numbers are severely kept down by the gun and the traps of the keeper, an immediate increase takes place in the event of any relaxation of the pressure. The Merlin breeds rather later than the other hawks, and the middle or end of May is quite soon enough for eggs. Its nest is simply a depression scratched in the soil, and filled with what moss and dead twigs happen to be close by, so as to make a little platform, and the sitting bird, which is very often the male, breaks off all the twigs from the heather growing within reach, and gathers them together to form a slight rim. The eggs are three or four in number, and being usually laid among tall heather, and hidden by overhanging branches, are not at all easy to discover. The food of this species consists mostly of beetles, mice, and small birds.

[On the Continent the Merlin frequently deposits its eggs in old nests built in trees, and it occasionally does so in the British Islands. The following, from Mr. Howard, is interesting, although the locality mentioned is in the next county :—" Mr. Morris, of Sedbergh, tells me that in June 1887, a Merlin took possession of the deserted nest of a Carrion-Crow placed at a height of six feet from the ground in a thorn-bush on Firbank Fell, near Kendal. The gamekeeper shot both old birds and took the young in Morris's presence."—Ed.]

GENUS TINNUNCULUS.

RED-FOOTED FALCON.

TINNUNCULUS VESPERTINUS (Linnæus).

A young male was obtained in Heaton Park in 1843, shot whilst feeding on dragon-flies over a " pit " there, and is now in the possession of Mr. Wright Johnson of Prestwich. In the museum of Peel Park, Salford, are a male and female which Mr. John Plant says he purchased from some birdstuffers named Harrop in 1850, who told him they had been shot in Prestwich. Clough about the year 1843.

KESTREL.

TINNUNCULUS ALAUDARIUS (J. F. Gmelin).

LOCAL NAMES—*Red Hawk, Windhover, Stannel.*

The Kestrel is the commonest of the birds of prey, and is resident throughout the year. It is universally distributed, and breeds wherever there are suitable woods, always showing a preference for those growing at some elevation. Not invariably, however, does it rear its young in trees; I have known it occupy for the purpose a ruined shooting-box among the hills, and very often a ledge of rock on some desolate moor is chosen, the soil being just scratched away a little, and the eggs laid on a collection of old "pellets." These pellets, which are almost entirely composed of the elytra of beetles and fur of mice, are also the only lining prepared when a nest in trees is occupied; and it is very probable that the birds roost in winter near where they intend to breed next spring, and that this lining represents the accumulations of the interval. Kestrels never build a nest for themselves, and always appropriate a Magpie's or Carrion-Crow's of a previous year, taking not the slightest trouble in repairing it. The full number of eggs is six, but only five are laid in many instances, and if the nest be robbed before the whole are deposited, the birds simply move to the nearest old one, perhaps not twenty yards away, and there the remainder are placed. The Kestrel begins to sit about ten days earlier than the Sparrow-Hawk; late April or early May being the most usual time, though fresh eggs may be taken from the middle of April to the beginning of June. In many districts a continual decrease is going on, owing to the efforts of the game-keepers, and when the young have to be fed, there is no doubt the parents are less particular, and that young Pheasants or Partridges occasionally fall a prey. Mr. W. A. Durnford relates (*Zool.*, 1878) how he disturbed a Kestrel which was carrying away a young Cuckoo, but items of this sort seldom form part of its regular pabulum. Dr. Skaife (*Mag. Nat. Hist.*, 1838) remarks on the evident good feeling which existed between a colony of Starlings and a few pairs of Kestrels which bred in the fissures and on the ledges of Alum Scar, near Blackburn, and how the usual bustling activity of the former became a precipitate flight for shelter whenever a Sparrow-Hawk appeared from the woods below. In autumn and winter the Kestrel frequents the sand-hills of the shore, and the low-lying moss-lands, feeding on the mice which abound there, and Mr. C. E. Reade tells me that he once shot a specimen at Urmston whose breast had been pierced by the sharp wheat-stubble, the result, no doubt, of some too reckless stoop upon its prey.

GENUS PANDION.

OSPREY.

PANDION HALIAËTUS (Linnæus).

An occasional visitor between October and May; few years pass without a specimen being seen or shot in one place or another. I have gathered a long list of occurrences during the last forty years, and there is no doubt many more have taken place, of which no record has been kept. Of the latest it may be said that one was shot on the Scarisbrick estate in April, 1880 (R. J. Howard), and in October, 1881, a fine male was killed on the Bowland fells, this being the second here within a short period. Mr. J. J. Hornby writes me that in the winter of 1881 also, an Osprey was seen about the lake at Knowsley for some weeks, and that orders were given that it should not be shot. This species is much more frequently seen about the inland rivers and lakes than upon the sea-coast. It was known to Dr. Leigh ("Nat. Hist. Lanc., &c.," 1700), for, speaking of "an Asper" which he figures, and which was killed upon the smaller Martin Meer, he says, "the Asper is a species of the Sea-Eagle, and is sometimes observed in these parts; its food is upon Fish, &c., &c. .... as to the oyl of the Asper, so vulgarly famed for alluring of fish, it is only a general mistake, and in no wise answers the end." The smaller Martin Meer of Leigh (which he says "is famous for Pearches, and vast quantities of Fowls, as Curlews, Curleyhilps, Wild Ducks, Wild Geese, and Swans, which are there sometimes in great numbers"), is the present Marton mere, near Poulton-le-Fylde, which drains into the Wyre; it is much reduced in size, covering now under ten acres probably, and becoming less every year.

ORDER STEGANOPODES.

FAMILY PELECANIDÆ.—GENUS PHALACROCORAX.

CORMORANT.

PHALACROCORAX CARBO (Linnæus).

LOCAL NAMES—*Scarth, Scarf.*

Owing to the entire absence from the Lancashire coast of any cliffs suitable for rock-loving birds, the present species is properly only a winter visitor. St. Bees Head, however, where, according to Dr. Parker, it breeds regularly, is not very far away, and odd birds, doubtless stragglers thence, are seen at all seasons. Byerley states (1856) that on the sand-banks near Liverpool it is very numerous, but, as a rule, only one or two birds are seen frequenting any one part of the coast in winter. In the Lune, Mr. T. Jackson tells me it may often be seen fishing, and, when tired, standing on a sand-bank, with wings spread out, for hours at a time, especially if the sun be shining. From its expertness in diving, it is seldom shot, and, naturally, it is seldom seen inland in this county, though it has occasionally been met with on the larger reservoirs.

SHAG.

PHALACROCORAX GRACULUS (Linnæus).

According to Dr. Parker, the Shag breeds at St. Bees Head in smaller numbers than the Common Cormorant

The Rev. H. A. Macpherson says that the Shag does not nest at St. Bees *now*, if it ever did.—Ed.]; it is very rarely seen near our shores. One was shot, Mr. J. B. Hodgkinson tells me, in the autumn of 1880, between Arnside and Silverdale, and Byerley says that it has been met with in several instances near Liverpool at the end of the year, after the breeding-season.

GENUS SULA.

GANNET.

Sula bassana (Linnæus).

A winter visitor, frequently appearing at sea in considerable numbers, and being blown inland during severe storms. Mr. John Weld writes me that on the 28th of November 1872, he came upon a fine specimen asleep in the middle of a field in Leagram. It had quite the appearance of a small drift of snow. He was able to steal up to it, and seize it by the neck, the head being completely buried in the scapular feathers. He carried it home with some difficulty, but it only survived its capture a few days. In October, 1878, also, an immature bird was taken on a night-line in the same neighbourhood.

ORDER HERODIONES.

FAMILY ARDEIDÆ.—GENUS ARDEA.

HERON.

ARDEA CINEREA, Linnæus.

LOCAL NAMES—*Crane, John Crane, Johnny, Johnny Gant, Long-neck, Jammy Long-neck, Jammy, Frank.*

Relics of the many Heronries which have existed within the last quarter of a century in different parts of the county, those at Scarisbrick near Southport, and at Ashton near Lancaster, still flourish. The latter at the present time [1885] contains twenty-five to thirty nests, and the former from eight to ten, this having been the invariable number for a great many years. At Claughton-on-Brock, near Garstang, there used to be a strong Heronry, but unfortunately some few years ago the wood had to be cut down, and the birds got dispersed. Mr. W. Fitzherbert Brockholes has good hopes, however, that it is established anew in another wood, where some young were safely hatched in the spring of 1884.* Mr. Chamberlain Starkie informs me that between 1800 and 1810 the Duke of Hamilton, who then lived at Ashton

* [Mr. Brockholes writes, in April 1892:—"In spite of all the protection I can give them, they do not increase in numbers, but dwindle back to four or six pairs before each nesting-time." The Rev. H. A. Macpherson tells me that there was a colony in Rond-sea wood not long ago, and that another, in the Rusland valley, still flourishes.—Ed.]

Hall, brought some Herons from Hamilton Palace, and had them in the paddocks. He has no evidence that any colony existed before then, so that the beginnings of this one would seem to have been from birds which had been some time under confinement. A few pairs may be found breeding together in several other localities, but whether from persecution, or from disinclination on the part of the regular tenants to admit fresh members into the community, a Heronry, if it increase at all, does so very slowly. The Heron is resident, and is well known everywhere, travelling, as it does, over such a wide extent of country in search of food: in winter it may be seen regularly on all the river estuaries, sometimes in flocks of thirty or forty individuals. Its eggs, varying from three to six in number, but averaging four or five, are laid from the middle of March to the middle of April, according to the character of the season. The nests are almost invariably placed at the extreme summit of a tall tree, preferably spruce or larch, on the branches a little way out from the trunk, and are built of large sticks, lined with finer twigs. Between the extreme edges, though the sticks here are very loosely put together, they measure about two feet and a half, and in the middle are six inches thick. Some nests have a considerable depression, preserved until the young are hatched, when of course they are flattened completely, whilst others are almost flat to begin with, and it often happens that eggs are blown out in high winds. This was the case during the great storm of March 1883, at the Heronry at Browsholme (near Clitheroe, but just over the Yorkshire border), which, since the birds returned to it (after a long absence) in 1877, has been under the constant observation of Mr. T. Altham and myself. The young are hatched at different ages, and incubation lasts about three weeks; the full plumage being assumed after pretty nearly the same length of time. The number of nests in this Heronry in 1877 was eight, and since then it has never exceeded sixteen.

PURPLE HERON.

Ardea purpurea, Linnæus.

[In *The Zoologist*, 1887, p. 432, Mr. J. Pickin records an adult male of this species, killed on April 7th of that year near Alderley Edge, about thirteen miles from Manchester.—Ed.

GENUS ARDETTA.

LITTLE BITTERN.

Ardetta minuta (Linnæus).

This species is very rare, and generally occurs in summer, though Byerley states that one was shot at Aigburth in 1854 in January. Bullock ("Companion to Liverpool Museum," 7th edit., 1809) says "one was killed perching on a tree near Manchester in June, 1808," and in a museum at Burnley is a specimen from Foulridge, Colne, taken in 1816. Mr. Thomas Webster records (*Zoologist* 1849, p. 2499) that on May 19, 1849, a Little Bittern was shot near the reservoirs of Gorton, and Mr. J. B. Hodgkinson informs me that specimens have occurred on Freckleton Marsh, and at Barton near Preston, and that one in his possession was shot at Crossens near Southport.

GENUS NYCTICORAX.

NIGHT-HERON.

Nycticorax griseus (Linnæus).

A fine specimen of the Night-Heron was shot on the marsh near Grange-over-Sands in May 1848 (Ant. Mason), and in Morris's *Naturalist*, 1853, Mr. James Bost says one is in his possession, killed near Blackpool on June 14, 1853, a description following, which shows it to have been in the plumage of the adult. Mr. Chamberlain Starkie writes me that he got a young bird at Ashton Hall in the spring of 1879. Byerley says ("Fauna of Liverpool," 1856) that he remembers two or three instances of the Night-Heron having been shot within the last twenty years.

GENUS BOTAURUS.

BITTERN.

Botaurus stellaris (Linnæus).

Local Names—*Bitter-bump, Bittery-bump.*

The Common Bittern is now only a winter visitor, but occurs so frequently as to make the list of its captures not worth detailing. Out of the score or so which I find recorded during the last forty years, almost all have been taken in December and January; and the 15th of April, when the last I have heard of was shot, in 1882, at Lytham, is an unusually late date. The Bittern is naturally most often seen upon the low-lying peat-mosses and marsh lands, and there is no doubt that, years ago, in their unreclaimed state, these were tenanted for the purposes of breeding. Mr. John Weld says that this bird was not so scarce about Leagram, when the county was less drained, and that it was often heard booming in the marshy lands at night. He remembers one or two specimens being obtained previous to 1840, but it is now never seen in the neighbourhood. Mr. R. J. Howard is informed that, sixty years ago, it was common on Martin Mere: no specimens, however, have been taken there for twenty years, and the last, which was shot about that time ago, is now preserved at Crossens. The feathers of the Bittern are in great repute amongst anglers, and Mr. R. Standen was told in 1878 by an ancient follower of the craft that the most "killing fly" he ever made was from a Bittery-bump's feathers, which had been shot about forty years before at St. Michael's-on-Wyre. This fact may be presented to the angling fraternity without its probably having any effect in making the species still more scarce, for the progress of agriculture will certainly prevent its ever again becoming common.

AMERICAN BITTERN.

BOTAURUS LENTIGINOSUS (Montagu).

A male of this species was reported by Mr. James Cooper (*Zoologist*, 1845, p. 1248) as having been killed near Fleetwood on the 8th of December, 1845. It is now in the Preston Museum.

FAMILY PLATALEIDÆ.—GENUS PLATALEA.

SPOONBILL.

PLATALEA LEUCORODIA, Linnæus.

A Spoonbill was shot on the Ribble in the year 1840 by a man named Bramley, and passed into the collection of Mr. J. B. Hodgkinson, who generously presented it, with all his other birds, to the Preston Museum, and there it still remains. The time of the year at which it occurred is not known.

FAMILY IBIDIDÆ.—GENUS PLEGADIS.

GLOSSY IBIS.

PLEGADIS FALCINELLUS (Linnæus).

An example of the Glossy Ibis is catalogued in "A Companion to the Liverpool Museum, at the house of William Bullock, Church Street" (6th edit., 1808), as having been shot near Liverpool, and is described as "of a dark olive-brown colour with green reflections." A second is mentioned by Montagu ("Dict. Brit. Birds," Supp., 1813) as also having been shot near Liverpool, "within these two or three years," and the plumage is particularized, showing it to have been an immature bird. A third was reported by Dr. Skaife, of Blackburn (who said it had occurred near Fleetwood), to Yarrell, and the latter acknowledged the information in the 3rd edition of his "British Birds," published in 1856. A fourth, also immature, is in the possession of

Mr. J. B. Hodgkinson, who recorded its capture in the *Zoologist* for 1874, and who tells me that it was shot on Marton Mere near Poulton, in the year 1859. Much has been written on the identity of this species with the Liver, the traditional bird from which the city of Liverpool is said to take its name, but Yarrell, I think, exhausts the subject when he remarks (*op. cit.* ii. pp. 605–6), " The arms of the town of Liverpool are comparatively modern, and seem to have no reference to the Ibis. The bird has been adopted in the arms of the Earl of Liverpool, and in a recent edition of ' Burke's Peerage ' is described as a Cormorant holding in the beak a bunch of sea-weed. In the Plantagenet Seal of Liverpool, which is believed to be of the time of King John, the bird has the appearance of a Dove bearing in its bill a sprig of olive, apparently intended to refer to the advantages that commerce would derive from peace."

ORDER ANSERES.

FAMILY ANATIDÆ.—GENUS ANSER.

GREY LAG-GOOSE.

ANSER CINEREUS, Meyer.

The various species of Geese are seldom seen on the coasts of Lancashire except at the seasons of migration, and they appear most frequently in October and March. During hard frosts, however, or after storms at sea, they have often been observed in the remaining winter months, and flocks have been occasionally seen flying northward as late as June, and southward as early as September. These flocks usually fly at so great a height that it is impossible to distinguish the species, and there is considerable difference of opinion as to whether the present or the Pink-footed Goose is of more common occurrence. The Grey Lag-Goose has been shot many times round Morecambe Bay, and also on the inland reservoirs, and no doubt passes through every winter. Mr. T. Jackson says that, from his station at Overton on the Lune, the Grey Geese always go north-east in spring, and south-west in autumn, in flocks both large and small, and generally flying either in a straight line or in the form of a wedge.

With reference to this and the Pink-footed Goose, the following notes by Mr. Howard should be consulted. He says: "From opposite the Naze, seawards, the Ribble had no definite channel forty or fifty years ago; the estuary was all sand and its general level several feet lower than it is at present. Then the number of Knots, Redshanks, Dunlins, Godwits, and other waders was many times greater than now, but not more than 30 or 40 Wild Geese visited the river. Gradually the mud began to accumulate: training walls were put down, and the channel was confined to the north side of the estuary. As the mud increased in depth, and the ground approached its present level and became covered, first with glass-wort (locally known as samphire) afterwards with creeping bent-grass (*Agrostis stolonifera*), a greater number of Geese came; now we have about 200 in an average season from the end of September until the end of April. Since 1863 about 4,000 acres of marshland have been enclosed and cultivated, so that the estuary is practically ruined as a resort for waders, but the Geese remain all the winter; and as a rule there is no appreciable difference in the numbers which frequent the estuary.* Occasionally, however, they seem to collect, until several times the ordinary number are seen:—on 28th February 1886, it was estimated that at least 2,000 were on the river, but they were very wild and would not allow approach within 300 yards. In addition to the birds which winter with us, gaggles pass over during the periods of migration; in autumn flying S.W., in spring N.E. The Geese feed chiefly during the day on the out marsh on creeping bent-grass—a nutritious grass which gives a very early bite, and which enables the marsh to carry and keep in good condition a large stock of cattle; but if much harassed they fly inland, especially when there is a 'light' moon, and feed on cultivated grasses, clover, young wheat, rotten potatoes, and grain on the inner marsh and moss land, seldom, however, venturing more

* [This reclaimed land is coloured *red* in the map.—Ed.]

than a mile or so from the shore. They 'flight' later than the Ducks, usually about 8 or 9 o'clock. Martin Mere, after it was drained, but before it was thoroughly reclaimed and cropped, was the chief inland feeding-ground. John Cookson *then* shot Geese above ten pounds in weight and with flesh-coloured bills (Grey Lags). In addition to the Geese which are shot, some are taken in 'ring' nets, but more—together with other wildfowl—are caught in strong steel traps which are set in pools left by the receding tide, as well as in any plashes of freshwater which may be on the arable land within the embankment. In April, one year, Richard Iddon found that a Goose had been caught by the bill; part of the upper mandible taken off in a line with the nostrils, was in the trap: the following November he caught a Goose, probably the same bird, with this portion of the bill missing.

"For several years I have taken much interest in the distribution of Grey Geese in Lancashire, and never missed an opportunity of seeing any birds killed in the county. The Ribble estuary is undoubtedly more frequented by Grey Geese than any part of the Lancashire coast: about the Lune estuary they seldom alight; and Mr. T. Jackson, of Overton, informs me that he cannot say what species pass over during the seasons of migration, for he has shot only one Goose in his life. *I have not yet succeeded in getting a sight of a Grey Lag killed in the county.*"—R. J. H.

Mr. Hugh P. Hornby, in a letter to Mr. Howard, states that in January 1891 he got in Preston the only true Grey Lag-Goose he ever possessed; said, and he believes correctly, to have been shot on the Ribble.—Ed.]

WHITE-FRONTED GOOSE.

ANSER ALBIFRONS (Scopoli).

A winter visitor, not common, but has been shot both on the coast and inland. The Editors of the *Naturalist's Scrap Book* say (pt. 14) that this species comes down to the marshes and river (Mersey) at night to feed, passing the day on the moss-lands.

PINK-FOOTED GOOSE.

ANSER BRACHYRHYNCHUS, Baillon.

A regular winter visitor, specimens being killed every year, and by the wildfowlers of the Ribble estuary it is considered to be much more plentiful than any other species. Mr. R. J. Howard tells me that, owing to the receding of the sea from the Southport side, Geese altogether are more numerous. I believe that a few Grey Lags appear with the Pink-footed. According to Mr. Richard Iddon, the Geese arrive about the first week in October, and remain until the last week in April or beginning of May.

BEAN-GOOSE.

ANSER SEGETUM (J. F. Gmelin).

The Bean-Goose is occasionally shot on the coast, and there is little doubt of its forming a part of the migratory flocks of Grey Geese which pass over the county in autumn and spring.

GENUS BERNICLA.

BRENT GOOSE.

BERNICLA BRENTA (Pallas).

The Brent Goose is stated by Mr. W. A. Durnford to be a regular winter visitor to the shores of Walney Island, and is not uncommon on some other parts of the coast. The late Rev. J. D. Banister used to be familiar with it at Pilling, and from this locality a specimen was received by Mr. Hugh P. Hornby on the 19th December, 1882.

[Mr. Howard says that the Brent Goose is seldom seen about the Ribble estuary. On November 18th, 1885, he received one, shot on Hesketh Marsh, from Richard Iddon, who did not know the bird.]

BARNACLE GOOSE.

BERNICLA LEUCOPSIS (Bechstein).

A winter visitor, occasionally seen inland, and in severe weather appearing sometimes on the coast in considerable numbers; many having been shot in 1878 and 1879 round the shores of Morecambe Bay. Yarrell quotes Selby as saying that it is sometimes abundant on the Lancashire coast, and from early times the Barnacle seems to have been well known in this district. Willughby states in his "Ornithology" (Ray, 1678) that it "frequents the sea-coasts of Lancashire in the winter time," and Dr. Leigh also ("Nat. Hist. Lanc., &c.," 1700) says that "the Barnacle is very common," but the plate which he gives, and which is labelled "The Barnacle or Anser Bassanus" has evidently been drawn from the Gannet. The old superstition that Barnacle Geese issued from a marine shell, and were not hatched from eggs in the usual way, has some local interest from the fact that the island on which stands the old Pile of Fouldrey, near Walney, was held to be a very favourite place for their propagation. I subjoin some extracts from "The Herball or General Historie of Plants" Gathered by John Gerarde, enlarged, &c., by Thomas Johnson, London, 1633, p. 1587, which the author states to be "the naked and bare truth, though unpolished," but which, nevertheless, is not accepted by Johnson, who subscribes to a quotation from the "Phytobasanos" of Fabius Columna, which runs, "Conchas vulgò Anatiferas, non esse fructus terrestres, neque ex iis Anates oriri; sed Balani marinæ speciem!"

Gerarde writes, "what our eyes have seene, and hands have touched, we shall declare. There is a small island in Lancashire called the Pile of Foulders,* wherein are found the broken pieces of old and bruised ships, some whereof have been cast thither by shipwracke, and also the trunks and bodies with the branches of old and rotten trees, cast up there likewise; whereon is found a certain spume or froth that in time breedeth unto certain shels, in shape like those of the Muskle, but sharper-pointed, and of a whitish colour: wherein is contained a thing in forme like a lace of silke finely woven as it were together, of a whitish colour, one end whereof is fastened unto the inside of

---

* Willughby (op. cit. supra) among remarkable breeding-places names "A noted island not far from Lancaster, called the Pile of Foudres, which divers sorts of Sea-fowl do yearly frequent, and breed there."

the shell, even as the fish of Oisters and Muskles are: the other end is made fast unto the belly of a rude masse or lumpe, which in time commeth to the shape and forme of a Bird: when it is perfectly formed the shell gapeth open, and the first thing that appeareth is the foresaid lace or string; next come the legs of the bird hanging out, and as it groweth greater it openeth the shell by degrees, til at length it is all come forth, and hangeth onely by the bill: in short space after it commeth to full maturitie, and falleth into the sea, where it gathereth feathers, and groweth to a fowle bigger than a Mallard, and lesser than a Goose, having blacke legs and bill or beake, and feathers blacke and white, spotted in such manner as is our Mag-Pie, called in some places a Pie-Annet, which the people of Lancashire call by no other name than a tree Goose: which place aforesaid, and all those parts adjoyning do so much abound therewith, that one of the best is bought for threepence. . . They spawne as it were in March and Aprill; the Geese are formed in May and June, and come to fulnesse of feathers in the moneth after."

The shell which was supposed to have such extraordinary properties is really a species of multivalve, the *Lepas anatifera* of Linnæus.

GENUS CYGNUS.

WHOOPER SWAN.

CYGNUS MUSICUS, Bechstein.

The Whooper is an occasional winter visitor, seldom appearing except in very severe weather, but then frequenting Windermere, Coniston, and other large pieces of water, as well as the estuaries of the rivers, and sometimes following these a considerable distance from their mouths. It occurs in both adult and immature plumage, and generally in only small flocks. Dr. Skaife (*Mag. Nat. Hist.*, 1838) says that between February 7 and 17 in 1838, the winter of which year was "dreadfully severe," four were shot out of a flock of twenty-seven, in various parts of the Ribble, and the Rev. J. D. Banister also wrote in his journal, under date of February 3rd, 1838, "Eight Swans (white) were seen on Pilling sands. . . . A great many Swans have been shot and taken alive in this neighbourhood, some were of a dusky gray, . . others gray head and neck but generally white. I have examined two specimens white excepting a portion of feathers on the forehead of a rusty colour. . . March 26. This day I sent a fine living Swan by coach to the Earl of Derby. . . This bird was caught near Wyre Water in the frost by a boy in February."

During the series of hard winters from 1878 to 1881, examples of this species were seen and shot on almost all parts of the coast and moss-lands.

BEWICK'S SWAN.

Cygnus bewicki, Yarrell.

A winter visitor, very rarely seen.

The account given of the occurrence of a flock of this species near Manchester by the late Mr. John Blackwall ("Researches in Zoology," 1834) is so interesting that I venture to transcribe it. He says: "About half-past eight on the morning of the 10th December, 1829, a flock of twenty-nine Swans, mistaken by many persons who saw them for Wild Geese, was observed flying over the township of Crumpsall at an elevation not exceeding fifty yards above the surface of the earth. They flew in a line, taking a northerly direction, and their loud calls, for they were very clamorous when on the wing, might be heard to a considerable distance. I afterwards learned that they alighted on an extensive reservoir, near Middleton, belonging to Messrs. Burton & Sons, calico-printers, where they were shot at; and an individual had one of its wings so severely injured that it was disabled from accompanying its companions in their retreat. A short time since, I had an opportunity of seeing this bird, which resembled the rest of the flock with which it had been associated, and found, as I had anticipated, that it was precisely similar to the small Swan preserved in the museum at Manchester, which, I should state, was purchased in the fish-market of that town, about five or six years ago . . . . . of the habits and manners of this species little could be ascertained from a brief inspection of a wounded individual; I may remark, however, that, when on the water, it had somewhat the air and appearance of a Goose, being almost wholly devoid of that grace and majesty by which the Mute Swan is so advantageously distinguished. It appeared to be a shy and timid bird, and could only be approached near by stratagem, when it intimated its apprehension by uttering its call. It carefully avoided the society of a Mute Swan which was on the same piece of water. On the 28th of February, 1830, at half-past ten A.M., seventy-three Swans, of the new species, were observed flying over Crumpsall in a south-easterly direction, at a considerable elevation.

They flew abreast, forming an extensive line, like those seen on the 10th of December, 1829; like them, too, they were mistaken for Wild Geese by most persons who saw them with whom I had an opportunity of conversing on the subject; but their superior dimensions, the whiteness of their plumage, their black feet, easily distinguished as they passed overhead, and their reiterated calls, which first directed my attention to them, were so strikingly characteristic, that skilful ornithologists could not be deceived with regard to the genus to which they belonged. That these birds were not Hoopers may be safely inferred from their great inferiority in point of size. . . . .
I am informed, that when the Wild Swans were shot at, near Middleton, on the 10th of December, 1829, one of them was so reluctant to abandon the bird which was wounded on that occasion, that it continued to fly about the spot for several hours after the rest of the flock had departed, and that, during this period, its mournful cry was heard almost incessantly. In consequence of the protracted disturbance caused by the persevering efforts of Messrs. Burton's workmen to secure its unfortunate companion, it was at last, however, compelled to withdraw, and was not seen again till the 23rd of March, when a Swan, supposed to be the same individual, made its appearance in the neighbourhood, flew several times round the reservoir in lofty circles, and ultimately descended to the wounded bird, with which, after a cordial greeting, it immediately paired. The newly arrived Swan, which proved to be a male bird, soon became accustomed to the presence of strangers; and, when I saw it, on the 4th of April, was even more familiar than its captive mate. As these birds were strongly attached to each other, and seemed to be perfectly reconciled to their situation, which, in many

respects, was an exceedingly favourable one, there was every reason to believe that a brood would be obtained from them. This expectation, however, was not destined to be realized. On the 13th of April, the male Swan, alarmed by some strange dogs which found their way to the reservoir, took flight and did not return; and on the 5th of September, in the same year, the female bird, whose injured wing had recovered its original vigour, quitted the scene of its misfortunes and was seen no more." In 1830, also, a correspondent of the *Magazine of Nat. Hist.* says that, on the 31st of January about thirty Swans appeared on Windermere, Coniston, and Esthwaite, and that several were killed; the measurements of one of these, however, show it to have been a Whooper. The only other Bewick's Swan I have come across is one in the possession of Mr. T. Altham, shot from a flock of twenty-four birds on Foulridge reservoir, near Colne, about the year 1857, on the 14th of March.

GENUS TADORNA.

COMMON SHELD-DUCK.

TADORNA CORNUTA (S. G. Gmelin).

The Sheld-Duck is a resident species, and is found breeding, in more or less numbers, on the sand-hills of the whole of the Lancashire coast. It almost disappears for a while in autumn, but in winter occurs in flocks of three or four to a dozen individuals, and odd birds often wander to the inland moors and mosses. It is nowhere so common as on the island of Walney, and the partial protection it meets with there prevents any decrease in its numbers, but on approaching the large watering-places further south, it is terribly disturbed by the parties of visitors which frequent them, and no expectation can be held of its long continuance in these localities. Still it keeps a precarious footing, and within the last few years nests have been found both north and south of the estuaries of the Lune and the Ribble, whilst on the Mersey above Liverpool a pair or two breed regularly. Formerly it appears to have been very abundant, and Donovan ("Nat. Hist. Brit. Birds," 1794–1819) speaks of it as being "found in vast quantities on several of our sea-coasts, and particularly about the rivers and lakes in Lancashire and Essex." Mr. W. A. Durnford, also, who has an extensive acquaintance with this species, in noticing (*Zool.*, 1877) a flock of about a hundred which he saw in Walney channel on January 13th, 1877, writes of them as only "a small remnant of the thousands which, within the memory of man, used to frequent the warrens in this locality"; and the late Dr. Skaife (*Mag. Nat. Hist.*, 1838) says that about the mouth of the Wyre, where now it is very infrequent, it bred regularly. The Sheld-Duck lays from nine to twelve eggs in old rabbit-burrows, and usually eight to ten feet deep in them, but Mr. Durnford records an instance (*Zool.*, 1880, p. 241) in which he found as many as sixteen eggs at a distance of only three feet from the hole-mouth. The young are hatched late in May, or early in June, and take to the water at a very early age, rapidly becoming expert in diving; while the female employs all the usual feints of pretended lameness, if suddenly disturbed among her brood on shore. The nests are difficult to find, and the birds are very shy and wary; the lighthouse keeper on Walney told the late

Mr. H. Durnford (*Zool.*, 1873) that "during the time the female is incubating, after feeding, she, in company with the male, flies to the neighbourhood of her nest, and after circling once or twice in the air over the spot, to see whether the coast is clear, flies straight into the hole without alighting on or touching the ground; the male (called 'Mallard' by the lighthouse keeper), after performing one or two more circles, flies off to his feeding quarters."

GENUS MARECA.

WIGEON.

MARECA PENELOPE (Linnæus).

LOCAL NAMES—*Lady Wigeon, Russianett.*

The Wigeon is a periodical visitant in winter, and a spell of severe weather will bring it on the coast in very large numbers. It is only occasionally seen inland, and is a wary bird, being difficult of approach even in a gunning-punt. Mr. T. Jackson informs me that it begins to arrive in the Lune estuary about the end of October, leaving again early in March, and that if the weather be hard, the birds may be counted by thousands. In the dusk of the evening they come to feed on the salt-marshes, and go away to the open sea again by break of day. Flocks are often seen in the Wyre, but Mr. R. J. Howard says that on the Ribble Wigeon are not so plentiful as formerly, and from the records of the decoy*

---

* Duck-decoying is carried on most successfully from about 9 to 10 in the morning, and 3 to 4 in the afternoon. The birds fly off at dusk, and return at day-break to sleep and enjoy them-

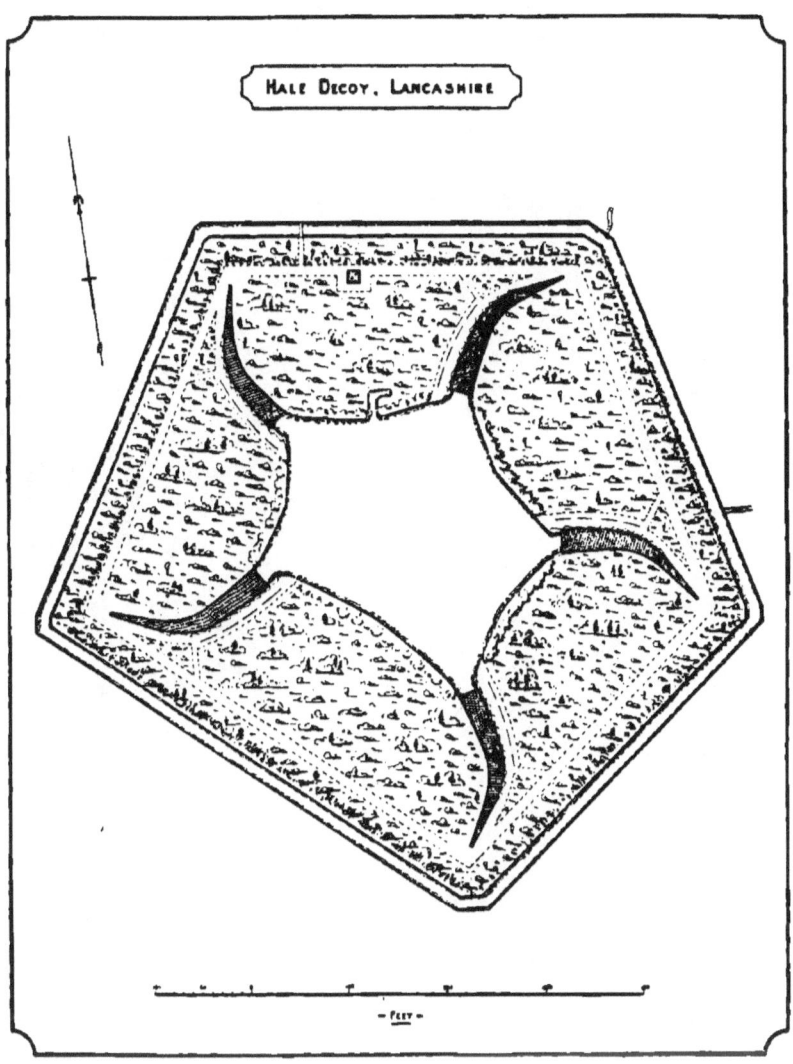

at Hale, it would appear that they have never been very common on the Mersey. This decoy is the only one now being worked in Lancashire, and through the kindness of Col. Ireland-Blackburne, I have been favoured with, as well as much other information, a copy of its records for the years in which they have been kept. I understand that before 1875 this was not done very carefully.

selves in the fancied security of the pool. The annexed plan is a correct drawing of the Hale decoy. The main pool has five arms or inlets called " pipes," curving away from it, so that it is not possible to see more than a short distance up them, and so arranged that whichever way the wind blows, one or other may be approached without getting to windward of the quick-scented wildfowl. These " pipes" are roofed over with netting, made as light as possible, and are gradually diminished in height and width till they terminate in a "tunnel-net." Wooden palings bound them, and on one side are built-obliquely, overlapping each other at regular distances and connected by low barriers, so that any one standing behind them is only visible to whatever is further up the " pipe," and cannot be seen by the occupants of the pool. Further aids to concealment, too, when approaching, are provided by continuous banks of earth and brushwood running parallel to the palings. The decoy-man, accompanied by his little dog, which should be as red and "foxy"-looking as possible, after ascertaining, by peeping through slits in the palings, near which " pipe" the fowl lie most conveniently, proceeds with the utmost caution till behind those nearest the entrance. Opening a little hole in one of the barriers, he slips the dog through, who, at once becoming visible to the fowl, trots up the bank inside till, arriving at another hole higher up, he returns through again, and so comes back to his master. The ducks, excited by curiosity, swim after the dog up the "pipe," and as soon as they are far enough, the man shows himself behind them, gesticulating violently but silently, and drives them in terror on and on, till the tunnel-net receives them, and this being loosened, and a twist given to it, they are quietly secured. All this has been done out of sight of any birds remaining on the pool, and the operation may be repeated till familiarity has begot contempt, and the appearance of the dog ceased to interest them.

Ducks Taken at Hale Decoy.

—	Wild Duck.	Wigeon.	Teal.	—	Wild Duck.	Wigeon.	Teal.
1801	34	1	78	1875	444	0	718
1802	28	3	222	1876	274	18	645
1803	28	23	221	1877	191	29	563
1804	90	24	241	1878	92	2	298
1805	92	0	137	1879	126	3	323
1806	73	0	334	1880	34	7	39
1807	33	4	65	1881	16	0	79
1808	50	16	196	1882	7	0	91
1809	26	3	130	1883	69	46	201
1810	82	5	91	1884	56	5	611
1811	57	11	73	1885	52	8	759
1812	278	39	71	And one Shoveler.			
1813	307	35	108				
1814	133	10	60				
1815	182	10	54	[Mr. R. J. Howard adds the following:—]			
1816	174	47	82				
1817	255	3	99	*1885-6	13	5	406
1818	220	13	43	1886-7	28	2	240
1819	181	5	4	1887-8	166	44	378
1820	227	3	2	1888-9	182	36	639
1821	77	8	9	1889-90	255	54	550
1822	414	54	86	1890-1	210	98	366
1823	45	15	19	1891-2	179	65	585
1824	52	25	61	Also two Pintails in 1887-8, five in 1888-9, and three in 1889-90.]			
1825	123	5	0				

There is conclusive evidence to prove that this decoy has been in existence for at any rate 150 years, but in 1754 it was much improved by the then Mr. Blackburne, who came to settle at Hale, from Orford Hall near Warrington, where there had long been a decoy, though since abandoned. The season lasts from September to the 15th of February, and the variations in the numbers caught in different years may be attributed to special and local causes rather than to an increase or decrease in the birds frequenting the locality.

[Mr. Howard sends the following:—"9th Jan. 1885.

* ["Col. Ireland-Blackburne writes me that on the 23rd January, 1886, they killed at one catch, 4 Mallard and 119 Teal, the largest catch ever chronicled at Hale."—R. J. H.]

A full-winged female Wigeon on reservoir in Blackburn Corporation Park; she seems quite at home with the pinioned birds. 7 June 1890. Full-winged Wigeon laid two eggs on island, on large reservoir in Corporation Park. These were taken by rats, I think—she has again nested under a laurel on the north of straight drive; this morning there are three eggs; 23 June 1890: disturbed, nest forsaken. 10 Dec. 1890. This bird which disappeared altogether about the above date turned up again to-day on the reservoir. 8 July 1891. Wigeon, pinioned bird, appeared on reservoir with seven young. Her nest was within two yards of one of the paths, under a rhododendron; all her eggs hatched. On the 10th the Park Supt. brought me a young one found dead. 30 Sept. 1891. Full-winged Wigeon, which has been away since midsummer, again appeared on Park reservoir. 21 Oct. 1891. Another full-winged Wigeon has come to the Park, a dark smallish bird, remarkably tame. It went into a large cage, used for catching the ducks, and was pinioned 20 Nov. Has since moulted into male plumage. Where can these full-winged birds have come from? Where does the duck go each summer?"]

GENUS DAFILA.

PINTAIL.

Dafila acuta (Linnæus).

Local Name—*Sea-Pheasant*.

A winter visitor, occurring only in small numbers, and seldom far from the sea. It has often been shot in September.

GENUS ANAS.

WILD DUCK or MALLARD.

ANAS BOSCAS, Linnæus.

LOCAL NAME—*Grey Duck.*

The common Wild Duck is a resident species, and is very generally distributed, breeding numerously in suitable swampy localities, though, owing to drainage, these are much fewer than they used to be. In winter it is seen on the coast in very large flocks, which, like those of all the other Ducks, are very much augmented in hard weather. Dr. Leigh (" Nat. Hist. Lanc., &c.," 1700) states that at Bold it was the practice then to feed the Wild Ducks in winter, and goes on to say, "great quantities of these birds breed in the summer season in pits and ponds within the demesne, which probably may entice them to make their visits in the winter; they oftentimes adventure to come into the moat near the Hall, which a person accustomed to feed them perceiving, he beats with a stone upon a hollow wood vessel, the Ducks answer to the sound, and come quite round him upon a hill adjoining to the water, he scatters corn amongst them, which they take with as much quietness and familiarity as tame ones; when fed they take their flight to the rivers, meers, and salt-marshes." The young of the Wild Duck are hatched in May, the eggs being laid early in April, sometimes late in March.

GENUS CHAULELASMUS.

GADWALL.

CHAULELASMUS STREPERUS (Linnæus).

A winter visitor, of rare occurrence. Mr. C. S. Gregson (*Proc. Hist. Soc. Lanc.*, &c., 1865-66, H. Ecroyd Smith) has one adult and one young bird, both shot near Altcar in April, 1865, and in May of the same year a pair were killed near Stretford (*Zool.*, 1865, C. W. Devis). One shot on Grimsargh reservoir in March 1860, was reported to the Preston Nat. Hist. Society by Mr. J. B. Hodgkinson, and Mr. Joseph Whitaker, of Rainworth, Mansfield, Notts, has been good enough to inform me that he has in his collection a male bird shot in Morecambe Bay in the winter of 1872-73. An adult male was shot December 15th, 1884, at Singleton, near Poulton-le-Fylde, and is now in the possession of Mr. T. H. Miller.

GENUS QUERQUEDULA.

GARGANEY.

QUERQUEDULA CIRCIA (Linnæus).

The Garganey is a very rare visitor, and only the following occurrences are known to me :—" One specimen at Ormskirk—Mather" (Byerley, " Fauna of Liverpool," 1856) : A male shot at Rufford, April, 1864, now in Mr. C. S. Gregson's collection (*Proc. Hist. Soc. Lanc.*, &c., 1865-66, H. Ecroyd Smith) : Two killed on Martin Mere about the year 1863 by W. Parker of Crossens (R. J. Howard).

COMMON TEAL.

Querquedula crecca (Linnæus).

Local Name—*Throstle-Teal.*

One of the earliest to arrive of the migrating Ducks, the Teal makes its appearance in autumn about the end of August or beginning of September, and at that time is met with in larger numbers even than the common Wild Duck, though as winter draws on this proportion is reversed. In hard frosts it is a frequent visitor to many of the inland reservoirs. A few pairs always remain throughout the year, and, on one or other of the "mosses" bordering the coast, its nest is found almost every season, but it is certainly decreasing as a breeding species. From the records of the decoy at Hale (p. 156), it will be seen that Teal have always formed a large proportion of the Ducks which have been captured there.

GENUS SPATULA.

SHOVELER.

Spatula clypeata (Linnæus).

Local Name—*Spoonbill.*

A winter visitor, present, no doubt, every season, but only in small numbers. Formerly it was much commoner, especially during the spring migration, and Mr. Robert Gray, who remarks, in his "Birds of the West of Scotland" (1871, p. 364), that he has "seen numbers of Shovelers shot on the Ribble, in Lancashire, early in

May," writes me under date November 11th, 1882, "In 1851 and 1852, along with the late Dr. Nelson, of Lytham, I often called upon a bird-stuffer in Preston, named Sharples, and it was on the occasion of these visits I had an opportunity of seeing the Shovelers and also Ruffs and Reeves *in quantities*. I have seen as many as twenty or thirty of each species in his hands at one time, all in the flesh. They had been shot on the banks of the Ribble, but I cannot now give the precise locality. Dr. Nelson, however, knew the place perfectly well, and had often shot both birds there himself." It is not infrequently met with by Snipe-shooters, and Mr. Hugh P. Hornby, who has several times killed specimens near St. Michael's-on-Wyre, tells me that at the Hale decoy it has been occasionally taken, but that he could not find any record of late years. An adult male was seen there on December 5th 1884, by Mr. R. J. Howard. Mr. J. B. Hodgkinson has had Shovelers, both male and female, shot in the breeding-season, and is confident that it sometimes nests on the hills in Higher Wyresdale, the characteristic blue feathers of the drake having been sent him in summer from birds shot in that district. A young bird was killed on Bury reservoir in August 1878, and preserved by Mr. R. Davenport, but the species is seldom seen so far inland.

GENUS FULIGULA.

TUFTED DUCK.

FULIGULA CRISTATA (Leach).

The Tufted Duck is not common, and is more frequently seen on the rivers and inland marshes than elsewhere. It has been shot in every month of the year except May and June, and birds killed in July are not likely to have been bred far away, though until last spring no information was forthcoming as to its nesting within the county limits. In June 1884, however, Mr. R. J. Howard tells me that a brood of nine young was hatched at Woodfold Park, and on the 25th of July, five of them, on their way from one lake to the other, were caught in a net, pinioned, and turned down again. The unpinioned birds left with the female, the male having disappeared some time previously, as soon as they could fly, she occasionally returning as if to induce her remaining young to accompany her: these, up to the 27th of September, were still very wild.

["On 20th July 1891, Tufted Duck hatched seven, and on 5th July 1892, eight young, in Blackburn Corporation Park."—R. J. H.]

SCAUP.

FULIGULA MARILA (Linnæus).

LOCAL NAME—*Blue-bill*.

A regular winter visitor, occasionally seen so early as September (*Zool.*, 1872, H. Durnford), though usually not till a month or two later. It is sometimes plentiful in Morecambe Bay, and Mr. T. Jackson tells me that large quantities are there taken in nets by the inhabitants of Arnside and Silverdale. It occurs in small flocks in all the river estuaries, but seldom leaves the sea far, frequenting mostly at such times the meadows flooded by the overflowing streams. [Since this was written several Scaup have been shot on the reservoir at Haslingden Moor, twenty miles inland.]

POCHARD.

FULIGULA FERINA (Linnæus).

A winter visitor, occurring periodically, and not uncommon. It is rare inland, and is not often seen at a greater distance from the sea than the flooded country near the coast: Mr. W. Naylor, however, preserved one killed at Whalley on November 12th 1867; Dr. Kershaw of Middleton has a specimen killed there in December 1879; and Mr. John Weld informs me that two males were shot on the 6th January 1880, on a small pond in Chipping. On the lake in Woodfold Park, near Blackburn, I am told by Mr. R. J. Howard, a pair of pinioned birds, which had been there four years without breeding, hatched five young (out of six eggs laid in May 1882), three of which flew away as soon as they were strong on the wing. These—for there can be little doubt they were the same individuals—returned in the spring of 1883, bred, and altogether about twenty birds were hatched. Most of the young were taken by the pike, and the original pair having been shot, about six pairs now remain, none of which are pinioned and all more than semi-wild.

[Mr. R. J. Howard sends me the following:—

"In 1886 a brood of hybrids between Pochard and Tufted Duck was produced in Woodfold Park. As the late Mr. Daniel Thwaites kindly allowed me to do what I liked with the birds, I sent one to Mr. H. Saunders, who, after exhibiting it at a meeting of the Zoological Society on December 21, deposited it in the British Museum (Natural History), two I forwarded to the Zoological Society's Gardens, and two females I put on the reservoir in the Blackburn Corporation Park. In 1887 another brood—the produce of the original pair—

was reared at Woodfold, but the female and the six young rambled to some densely reeded reservoirs near, and were not seen after 1st August. I purchased another female Tufted Duck in April 1888, but although the male Pochard (which had left for the winter) turned up, the birds did not breed. On June 5, 1889, the male Pochard again appeared, but stayed only a few days. I was determined, if possible, to have another brood of hybrids, and therefore purchased a pinioned male Pochard on February 20, 1890. The full-winged male Pochard came to the reservoir, but did not stay: the pinioned birds apparently paired. Both birds disappeared; neither Pochard nor Tufted Duck breed now at Woodfold. The female hybrids in the Corporation Park laid in 1889 and 1890, but the nests were taken. In June 1891 the birds were more successful and hatched one and six young respectively; four of these—all males—were reared, but two only were caught and pinioned. Of the unpinioned birds I shot one on August 25, 1891, the other flew away in December. The pinioned birds cannot now be distinguished from adult male Pochard. This year (1892) two more broods of hybrids were hatched : one of four on May 29, and the other of eight young on June 1. I believe that a Pochard is the father of all the hybrids. On June 11, a Pochard in the Corporation Park hatched six young ; when in down these are not to be distinguished from the hybrids."*—Ed.]

* [Pochard, 7 days old :—Centre of forehead, crown, hind neck and upper parts generally, olive-brown ; chin and throat, yellow ; broad streak over eyes ; sides of face, wing and rump patches, breast and belly, yellowish-buff, darker towards vent ; irides, yellowish-brown ; bill, dark liver-colour ; length, 7.5 inches ; width at widest part, one-third from tip, 4 inches ; sides of tarsi and toes, light olive ; front of tarsi and toes, and webs, dark olive..—R. J. H.]

GENUS NYROCA.

WHITE-EYED DUCK.

NYROCA FERRUGINEA (J. F. Gmelin).

Byerley ("Fauna of Liverpool," 1856) states that a specimen of this rare Duck was killed at Weston, which is on the Cheshire side of the Mersey, near Runcorn, in January 1854.

GENUS CLANGULA.

GOLDENEYE.

CLANGULA GLAUCION (Linnæus).

LOCAL NAME—*Mussel-cracker.*

A winter visitor, occurring regularly, and in hard weather frequenting the bays and estuaries of the coast in considerable numbers, as well as many of the inland pools and tarns. On Esthwaite water Mr. R. J. Howard says it is occasionally seen in small flocks, and Mr. W. A. Durnford (*Zool.*, 1876) records that "a small flock varying from two to twelve occupied, the whole winter of 1875, a large reservoir close to the Iron and Steel Works, Barrow. They were very persistent in keeping near the middle, and so escaped scathless. The last left about the middle of March 1876." This species seldom arrives in any numbers before the latter part of October.

GENUS HARELDA.

LONG-TAILED DUCK.

HARELDA GLACIALIS (Linnæus).

A rare winter visitor. Seen at Pilling (Rev. J. D. Banister, *Mss.*) : One, shot in winter of 1855 on Rishton reservoir (W. Naylor) : An old female, killed at Bury reservoir, December 1868, by F. Oates, seen in flesh (R. Davenport) : A young male, shot near St. Michael's-on-Wyre, October 26, 1882 (H. P. Hornby).

GENUS SOMATERIA.

EIDER DUCK.

SOMATERIA MOLLISSIMA (Linnæus).

A winter visitor, occurring rarely, in stormy weather. Mr. H. Miller tells me that, in the autumn of 1882, a flock of eleven birds was shot at close to Fleetwood, and three of them killed.

GENUS ŒDEMIA.

COMMON SCOTER.

ŒDEMIA NIGRA (Linnæus).

LOCAL NAMES—*Douker, Black Dyker.*

The Common Scoter is by far the most abundant of the Ducks which frequent the Lancashire coast, and is sometimes seen in Morecambe Bay in flocks of many thousands. The birds begin to arrive early in July, and Mr. E. C. Buxton (*Zool.* [1860], p. 7172) remarks that on the 7th July, 1860, he saw flocks at the mouth of the Ribble which must have numbered a thousand individuals, there being several Velvet Scoters mingled with them. Mr. T. Gough (*Zool.* [1848], p. 2230) also says that in the first week of July, 1848, fourteen

PIEL CASTLE AND DOUKER-NETS.

Common Scoters were met with on Windermere near Wray Castle, and that "this species occurs every year upon the lake, about the same time of the season, but never stays more than a day or two." In April and May the spring departure takes place, but odd birds may be seen along the coast the year through. Away from the sea, the Common Scoter is occasionally shot, but is rare. In the severe winter of 1879-80, however, Mr. R. Davenport tells me that twos and threes appeared every week on the Bury reservoir, where also, in 1878, a specimen was killed so late as the month of May. The Scoter is a shy and wary bird, difficult of approach even upon its first arrival, but, like the Scaup, it is sometimes taken in large numbers in the Donker-nets which are stretched for the purpose on many parts of the shores of Furness. These nets are of various lengths, but mostly about four feet wide, with a mesh of four inches. They are set on the sands near where the birds have been feeding the previous tide, this being evidenced by the droppings they leave, and the holes bored by them in their search for cockles and other small molluscs. Four small stakes are driven into the sand, leaving about fifteen inches visible, and the net is hung loosely between them, one stake at each corner. When the tide rises, and the Ducks come with it, whether they dive head-foremost into the nets, or get fast in them from beneath, they are rapidly drowned, and half a cart-load is not considered a very extraordinary day's catch.

VELVET SCOTER.

(Œdemia fusca (Linnæus).

The Velvet Scoter is sometimes represented by a few individuals among the flocks of the common species (p. 167), but it is of very infrequent occurrence, and only two instances of its being shot away from the sea have come under my notice. One of these was on Windermere, on May 23rd, 1848, the bird being a male, and Mr. Thomas Gough of Kendal, who recorded the capture (*Zool.* 1848, p. 2230), states that "the female was also observed about the same time"; the other is in the collection of Dr. Kershaw, and was obtained at Rhodes, Middleton, on November 20th, 1883. Mr. W. E. Beckwith has a young male, killed at Flookburgh on Cartmel sands in December, 1876, and Mr. S. Smith says (*Naturalist*, 1865) that a male in perfect plumage was shot by Robert Croft on November 20th, about two miles from the mouth of the Wyre, the gizzard being found to contain five marine shells (*Littorina littorea*), besides some young crabs. Other specimens have been obtained in the neighbourhood of Southport and Liverpool. [I saw several Velvet Scoters between Southport and Blackpool in September 1883.—Ed.]

SURF-SCOTER.

ŒDEMIA PERSPICILLATA (Linnæus).

Mr. Richard H. Thompson, of Lytham, writes me under date December 16, 1883 : " In December last I shot a very good specimen of the female Surf-Scoter (*Œdemia perspicillata*) about five hundred yards from the shore opposite this place. Mr. A. G. More pronounced the bird to be of the species." No other instance of the occurrence of this rare Duck on the Lancashire coast is upon record. [The Rev. H. A. Macpherson informs me that the above specimen is really a young male.—Ed.]

GENUS MERGUS.

GOOSANDER.

MERGUS MERGANSER, Linnæus.

In severe weather the Goosander follows the courses of the various rivers, especially the Ribble and its tributaries, and on them, and at Foulridge and the other reservoirs, many specimens, usually females or immature males, have fallen to the gun. It sometimes occurs in small flocks, and on the 22nd January, 1881, two females were shot on the Ribble, near Clitheroe, out of nine birds which, for above a week, had frequented the locality. It must, however, be considered as a winter visitor in only small numbers, and, although Mr. T. Jackson says he sees it on the Lune almost every season, it rarely visits the other portions of the coast. The female Goosander is probably the bird which Dr. Leigh (" Nat. Hist. Lanc., &c.," 1700) calls the Sparling-fisher, and which he says " is about the bigness of a Duck, and by a wonderful activity in diving catches its prey, and yields a very pleasant diversion when pursued by water-dogs." Willughby (" Ornithology," Ray, 1678) gives Dun Diver and Sparlin-fowl as synonyms for the female Goosander, and Sparling is stated by Pennant to be a name used for the smelt in Wales and the north of England, a fish which, he says, " inhabits the seas that wash these islands the whole year." Mr. T. Altham, however, tells me that on Morecambe Bay he has heard the Red-throated Diver called Sperlin-hunter.

RED-BREASTED MERGANSER.

Mergus serrator, Linnæus.

The late Dr. Skaife, in recording (*Mag. Nat. Hist.*, 1838) the capture of a " splendid male " near Southport on February 10th, 1838, says, " so rare is this bird in these parts that none of the bird-stuffers, nor the oldest sportsmen and fishermen, ever remember to have seen one of this species before." Mature specimens have, no doubt, at all times been rare, but it is probable that young birds have always been regular visitants, and Mr. W. A. Durnford says ("Birds of Walney," 1883) that this is so on the coast of Furness, and that they are generally called Goosanders, a mistake which may easily be made. On the Lune, Mr. T. Jackson says the Red-breasted Merganser is by no means rare, and that hardly a winter passes without its being seen: in that of 1880 six birds were killed there by one shot. Inland it is very seldom observed, and the only instance I have heard of has been reported to me by Mr. W. Naylor, who preserved one shot at Hacking boat on the Ribble in the winter of 1876.

SMEW.

Mergus albellus, Linnæus.

The Smew is a very rare winter visitor, and is generally seen only in hard weather *e.g.* 1891'. At Pilling it used to occur, according to the Rev. J. D. Banister (*Mss.*), and Mr. J. B. Hodgkinson says one was shot about the year 1874 near Preston. The origin of the two specimens noted by Dr. Skaife (*Mag. Nat. Hist.*, 1838) as having been obtained by him in the same town, is not stated. Mr. R. Davenport writes me that "one was shot on Bury reservoir some time ago, and is now in Mr. Johnson's collection at Radcliffe." To these Mr. R. J. Howard adds:—"One on the Wyre, in January 1886."—Ed.]

ORDER COLUMBÆ.

FAMILY COLUMBIDÆ.—GENUS COLUMBA.

RING-DOVE.

COLUMBA PALUMBUS, Linnæus.

LOCAL NAMES—*Wood-Pigeon, Stock-Dove, Cushie, Cushat* (often pronounced *Cowshot*).

The Ring-Dove is a resident species, and is universally distributed; becoming very common wherever there are suitable woods, and breeding numerously. In the grain districts it is very destructive, coming in flocks, often of thousands, and feeding in spring on the newly-sown fields, and in autumn on the ripe beans and other crops: Mr. Hugh P. Hornby, in a specimen killed on October 30, 1872, found as many as eighty beans. It is an early breeder, and the first clutch of eggs—for two or three broods are hatched during the season—is generally laid in the beginning of April. The number of eggs is two (very rarely one or three), and incubation, the duties of which are shared by both birds, sometimes commences immediately the first egg is laid, young of different sizes being not uncommonly found in the same nest. The nest is never placed at a very great height, and is usually a very slovenly structure, but in some cases considerable care is taken in forming a neat depression in the sticks which compose it, and in lining this with finer twigs. When the Ring-Dove is disturbed whilst sitting, or after the hatching of the young, it almost invariably flies first to the level of the ground on leaving the nest, thence curving upwards again when some distance has been attained; this is probably a primitive attempt at feigning lameness, which H. W. F., in the *Field* of July 5th, 1873, says he has observed as a trait of this species.

STOCK-DOVE.

Columba œnas, Linnæus.

Local Names—*Rock-Dove, Hill-Pigeon, Sand-Rock.*

In his paper in the *Ibis* of 1865, Mr. A. G. More writes that he is informed by Mr. J. F. Brockholes that the Stock-Dove breeds regularly in South Lancashire in fir-trees and ivy. With this exception, and perhaps also the higher part of the Wyre valley, up to 1877, it was only known on the coast, at least as a breeding species, but about that year a remarkable extension of its range took place. It then appeared on the banks of the Ribble at Balderstone, in various places on the Hodder, at Hapton Scouts, and on Pendle Hill, and has since been seen every year in these or neighbouring localities, and, where undisturbed, in steadily increasing numbers. Mr. J. J. Hornby took a nest in 1878 in Lower Wyresdale, Mr. John Weld says that a pair have bred for some years on the Greystonley brook in Bowland, and at Billinge, near Blackburn, Mr. R. J. Howard writes me that it now nests every year. From Walney, and the sand-hills on the neighbouring mainland, along the whole range of coast to near Liverpool, though nowhere common, it has always been well known, breeding in the rabbit-holes so numerous there. Inland also a rabbit-burrow is a very favourite nesting situation, varied occasionally by a cleft in some precipitous bank, and often under the shelter of a tree-root. Here, about arm's-length from the mouth, the two eggs are laid on a few sticks gathered together, and fresh ones may be found from so early as the first week in March to the first week in May. Two sets of young are brought up each spring, and in one instance Mr. T. Altham found two fresh eggs, in the same hole as, and close to, the nearly fledged young of the former hatching. In winter, owing to its flocking with the Ring-Dove, it is not so often observed, and is probably more numerous than is supposed. Mr. Hugh P. Hornby writes me that whilst out at dusk trying for Wood-Pigeons at Winmarleigh on January 8th, 1884, he killed four birds from separate flocks of a dozen or fifteen which flew past him, and these all proved to be Stock-Doves.

ROCK-DOVE.

COLUMBA LIVIA, Bonnaterre.

Much confusion exists locally as to the present species, owing to its name being used in many places to represent the Stock-Dove, and matters are not mended by the fact that Stock-Dove is the name almost invariably used for the Ring-Dove. The Rock-Dove is of rare occurrence in Lancashire, and this is not to be wondered at when its preference for a rocky and cave-indented sea-coast is considered. It breeds, however, at Whitbarrow Scar, in Westmorland, just over the border, and Mr. T. Jackson says that he sees it near Overton with the Ring-Doves, and that it occasionally breeds in an old quarry that has been closed for some time. The Rev. H. A. Macpherson tells me that all the available evidence goes to show that the Whitbarrow birds are merely Stock-Doves.—Ed. Mr. R. Standen has taken eggs which he believes to be of this species from Langden Fell, where he says it nests in rocky crannies in almost inaccessible situations, and such a position is certainly very likely ; but so many cases have occurred of the domestic Pigeon reverting to a feral state, that careful identification is necessary in all instances, and indeed a strong opinion is held by many naturalists that the true Rock-Dove never nests inland except in *caves*.—Ed.]

GENUS TURTUR.

TURTLE-DOVE.

TURTUR COMMUNIS, Selby.

The Turtle-Dove is a rare and occasional summer visitor, and no instance of its nesting within the county limits is on record. Examples have been shot in many localities, mostly in May and June, August and September, and these have no doubt been on migration.

Mr. R. J. Howard informs me that the "Turtle-Doves" recorded as breeding at Mytton Hall (Rev. J. Gerard, *Stonyhurst Magazine*, 1887) proved to be examples of the Oriental species, recently turned out of their cage.—Ed.]

ORDER PTEROCLETES.

FAMILY PTEROCLIDÆ.—GENUS SYRRHAPTES.

PALLAS'S SAND-GROUSE.

SYRRHAPTES PARADOXUS (Pallas).

This Asiatic species, in its extraordinary invasion of Europe in 1863, appeared in Lancashire only one day later than in any other part of England. Two males and one female were shot on the 21st of May, out of a flock of fourteen, at Thropton, in Northumberland (Yarrell's "Brit. Birds," 4th ed., Saunders), and on the 22nd, a covey of about fourteen was reported by Mr. E. J. Schollick in a letter to the *Times* of May 26th; this Professor Newton (*Ibis*, 1864, p. 210) says he believes to be the first published notice of the arrival of Syrrhaptes in England. Mr. Schollick wrote that the birds were seen on the Isle of Walney, and that his informant, who had just shot a beautiful brace, a cock and hen, told him they were very tame, permitting a near approach while feeding in a field of corn, and rising with a peculiar cry, but not flying far. In the *Zoologist* for 1863, Mr. T. H. Allis, of York, reported several others which were obtained in Morecambe Bay, near Lancaster, from the last week in May to the end of June, stating that the ovaries of the females contained eggs in various stages of development. No further examples appear to have been noticed until one was shot at Risley (*Nat. Scrap Book*, part ii. 1864,

C. S. Gregson), "which seemed to have lived near a town for some time, being as dirty as a Liverpool Sparrow," presumably the same specimen recorded by Mr. Gregson (*Proc. Hist. Soc. Lanc., &c.*, 1865-66) as "a male, shot near Warrington December 25th, 1863." Mr. R. J. Howard tells me that he has examined two males, which were shot by a gamekeeper on the Scarisbrick estate, in a field of spring oats, in 1863: they were the only birds killed out of a flock of about fifteen, the remainder flying off in a northerly direction.

[The following extracts are from a paper contributed to *The Zoologist* for 1889, by Mr. R. J. Howard:—

"In 1888, as in 1863, Lancashire was favoured with a visit from this interesting species; the second invasion, however, was on a larger scale, with a correspondingly heavier death-roll than that of twenty-five years ago. The first arrival of the birds, at almost the same date as in 1863 (as regards Lancashire, within two days of the same date), is remarkable. In this report I propose to deal with occurrences which have come within my own knowledge in the county of Lancashire, excepting the Furness district, which the Rev. H. A. Macpherson has included in his report on the subject, published in the 'Transactions of the Cumberland and Westmoreland Scientific Association.'

"May 20th.—Eight were seen to alight on the moss about a mile north of St. Michael's-on-Wyre. On the following morning Cuthbert Baines, a farm-labourer, shot four of them (two males and two females); the rest flew N.W. These are the birds referred to by Mr. Hugh P. Hornby (*Field*, June 2nd), who writes me that he was misinformed as to the number of birds in the flock, and the date when his were shot, and that his notes are consequently incorrect on these points.

"May 25th.—One seen flying across Tarleton Moss, against a strong east wind, by Henry Cookson.

"June 1st.—Two males and one female shot out of a flock of seven on St. Michael's Moss, by C. Baines. One male and one female in Mr. Francis Nicholson's collection; one male in Mr. W. B. Wardle's collection.

"June 2nd.—A flock of about twenty seen, by a party of Pigeon-shooters on the Manchester Racecourse, flying from the direction of Trafford Park; the birds, after passing over the course, wheeled and returned to the park.

"June 3rd.—Three males shot out of a flock of twenty on Rawcliffe Moss, by John Taylor; the remainder of the birds flew N.W., and were probably the seventeen which the Rev. H. A. Macpherson says were seen on June 11th on the north end of Walney. One in Mr. F. Nicholson's collection; one in the Blackburn Museum; one in my own collection.

"June 7th.—A solitary female shot on St. Michael's Moss, by C. Baines. In my own collection.

"June 30th.—One seen near Blackstone Edge Reservoir, by James Sutcliffe, gamekeeper.

"September 3rd.—One seen on Tarleton Moss, by Henry Cookson. This bird was flying west along the same line as the one which he saw on May 25th; but it was travelling in the opposite direction.

"From the above list it appears that in Lancashire fifty-nine Sand-Grouse have been seen and eleven (seven males and four females) killed. It is quite possible, however, that the flock of twenty seen near Manchester on June 2nd was the same as that observed on Rawcliffe Moss, forty miles N.W., on the following day, for most of the birds which escaped the gun flew off in a north-westerly direction. All those birds referred to, with the exception of that seen on Blackstone Edge, were met with in the low-lying district of West Lancashire, chiefly on the moss-land. Those seen near St. Michael's were partial to oat-fields, and were seldom, if ever, observed on the old grass-land. Cuthbert Baines told me that the birds were wild, and would not allow him to approach within 150 yards in the open; he had to creep down the moss-ditches to get within shot. The birds rose quickly the instant his head appeared above the edge of the ditch, and would not permit him to take the 'pot' shot invariably adopted with Dotterel. After being flushed, whether shot at or not, they usually flew a few hundred yards and returned in a short time to the same field; in this respect, as well as in their partiality for oat-fields, resembling Dotterel. They do not carry away much shot; all were killed with No. 10 at about thirty yards' distance.

"These Sand-Grouse would, I think, have little difficulty in finding an abundant supply of suitable food on our moss-land. In addition to grain (any kind of which it appears the Sand-Grouse will eat), most of the moss-land is full of the seeds of goose-foot and various species of knotgrass (*Polygonum*); seeds of the latter, with germinating power unimpaired, are found buried several feet in the peat, and are constantly being brought to the surface as the land is worked. Seeds of the goose-foot (*Chenopodium album*), a very common weed, were found in the crops of the Lancashire-killed specimens, and it appears that the seeds of a nearly-allied plant, *Agriophyllum gobicum*, formed the bulk of the food of the Sand-Grouse in Central Asia. Six of the birds killed at St. Michael's have passed, in the flesh, through my hands; and the contents of the crop of the other were sent to me by Mr. Nicholson. I forwarded the crops and gizzards to Mr. Robert Holland, Frodsham, who very kindly furnished me with the following particulars:—

"1. Crop: red clover, a few seeds of Italian rye-grass, and knotgrass (*Polygonum persicaria* or *lapathifolium*). Gizzard: half the bulk, small fragments of white quartz; seeds, knotgrass, red clover, and alsyke.

"2. Crop: red clover, a few seeds of Italian rye-grass and knotgrass.

"3. Crop: same as No. 2, with a few seeds of mouse-eared chickweed.

"4. Crop: knotgrass and red clover, a few seeds of trefoil, Italian rye-grass, perennial rye-grass and meadow-fescue. Gizzard: five-sixths of bulk, small fragments of white quartz; seeds, knotgrass, goose-foot, alsyke, and Italian rye-grass.

"5. Crop: red clover, Italian rye-grass, knotgrass, and goose-foot.

"6. Crop: same as No. 5.

"7. Crop: knotgrass, goose-foot, mouse-eared chickweed, Italian rye-grass. Gizzard: one-third of bulk, white quartz; seeds same as crop.

Plumage, Dimensions, Weight, &c.

		Length.	Wing.	Central Tail-feathers.	Weight.
1.	Male	16·9	9·4	7·6	9½ ozs.
2.	Female	15·1	8·2	5·2	10½ „
3.	Male.				
4.	„	15·2	8·7	6·2	9¾ „
5.	„	16·1	9·1	7·5	9¾ „
6.	„	16·3	8·55	6·3	9 „
7.	Female	13·0	8·3	4·0	9 „

"The birds were in fair condition; the female, No. 2, was very fat: hence her weight, for she had little in her crop. The eggs in this bird and No. 7 were about the size of No. 4 shot. The testes in the male were well developed; in No. 1 the left testicle was ·48 × ·22, right ·42 × ·3; in No. 5 the left was ·42 × ·32, right ·26 × ·26. The plumage was clean, though bleached and worn. The birds which passed through my hands had cast a few of the inner primaries and the secondaries, giving the wing a very peculiar indentation. In No. 1, the new primaries (the ninth and tenth) project about one inch beyond the coverts, are lavender along each side of the shaft, gradually shading to black towards the edges and tips, the edges rich buff ·2 wide; the new secondaries rich buff, with black stripe ·3 in width, along outer web, leaving a narrow border of buff; one of the central rectrices, new, 4 in. long. No. 4 has the colours the brightest of any I have seen. Abdominal band rich velvety black; pencillings of chest-band very clear; three inner primaries moulted. No. 7, female, is the most forward in moult. A few scapulars, one of the elongated tail-feathers, 3·8 in. in length; the secondaries, and the three inner primaries with their coverts, new; the eighth and ninth primaries almost full-grown; the tenth is hidden by the coverts. The black on the new primaries, not so well defined in outline as in those of the males, giving the centre of these feathers a mottled appearance. Abdominal band dark umber; gular band distinct; no trace of chest-band. This is the only bird which shows any new body-feathers."

As regards the Furness district above mentioned (*supra*, p. 177), the Rev. H. A. Macpherson wrote the following:—"On the present occasion Mr. W. Duckworth visited Walney on June 4th (1888) and subsequently. He ascertained on that date the presence of a flock of fourteen, and another of seven, the birds having arrived on May 19th. Between that date and June 18th, seven were shot and sent to a taxidermist at Barrow. On June 11th a flock of forty, and another of seventeen, appeared at the north end of Walney; and on June 17th a flock of eight were seen at the south end of the island."—Ed.

ORDER GALLINÆ.

FAMILY PHASIANIDÆ.—GENUS PHASIANUS.

PHEASANT.

Phasianus colchicus, Linnæus.

Resident, and numerous where preserved; artificial feeding being especially resorted to away from the grain-growing districts. So much crossing has taken place with the Ring-necked Pheasant of Southern China (*P. torquatus*), that specimens showing more or less of a white collar are much commoner now than those without. From ten to seventeen eggs are laid from April to May, and I have seen nests among the heather on the top of Longridge Fell, 1,016 feet above the sea, though generally they are on the lower ground.

GENUS PERDIX.

PARTRIDGE.

Perdix cinerea, Latham.

The Partridge is a common resident, and is very generally dispersed over the whole county. It varies very much in numbers with the wetness or dryness of the breeding-season, and hard shooting and the ceasing of corn-growing have in many districts made it comparatively scarce. It lays from ten to seventeen eggs, generally a week later than the Pheasant, and commencing the middle of April. Sometimes two females will lay in the same nest, and as many as twenty-three young have been hatched together.

GENUS COTURNIX.

QUAIL.

COTURNIX COMMUNIS, Bonnaterre.

Once a common summer visitor, the Quail has, within but a few years, become exceedingly scarce, and indeed, for the last three or four seasons, hardly a single bird has been seen. Except one example in September, 1884, it has not been noticed at St. Michael's-on-Wyre since 1874, in 1878 it disappeared from Tarleton Moss, and from many other localities in the Fylde and neighbouring districts, where a little while ago it was common, I have it reported as absent almost altogether. A general diminution in its numbers has, no doubt, been going on for a long time. Dr. Leigh ("Nat. Hist. Lanc., &c.," 1700) avers that it was common in his day, and the late Rev. J. D. Banister, writing in his journal (which, owing to the kindness of his son, the Rev. E. D. Banister, I have had an opportunity of perusing), under date December 9th, 1836, says, "I saw a fine Quail on the edge of Pilling Moss. Do they migrate? It is stated here by persons that formerly they were as numerous as Larks, but were destroyed by a very severe frost about sixty or seventy years ago and were never plentiful since." At St. Michael's-on-Wyre a careful record has been kept since 1865 by Mr. Hugh P. Hornby of the Quails shot there in autumn, and from this it appears that, whereas in 1865 and 1866 there were killed respectively twenty and twenty-six birds, in no year since has the number exceeded nine, and, as stated above, since 1874 only one bird has been seen altogether. The causes of this seem rather obscure, but I believe that the practice of mowing grass by machines, which shave so close to the ground that the sitting birds are invariably destroyed, is an important one, and I am confirmed in this opinion by Mr. T. Jackson, who says that he has killed many single ones in this way, and that once he took off the heads of three in one day. Its eggs have been taken over the whole of the Fylde, and Mr. R. Davenport tells me that at Unsworth, near Bury, it bred annually for a long time. The nest has often been found in the neighbourhood of Southport, and C. P. A., a correspondent of the *Field* of September 28, 1867, states that, presumably in the year named, a brood of seven young was reared on the Bickershaw estates, near Wigan, being the first time the species had been known to breed in that locality. In Furness it is rare, though here also it is stated to have once been not uncommon ("Birds of Walney," 1883, W. A. Durnford). It is usually shot in greatest numbers in September and October, but has often occurred up to December and January, and it is very likely that some individuals remain the whole winter in the vicinity of the breeding-ground. When disturbed by shooting-parties, the Quail, which sits very close, rises singly, and not in bevies, and Mr. Hornby has only met with one instance in exception to this, when ten birds once rose together. The eggs appear to be laid in May.

[Mr. R. J. Howard writes:—" During 1885 the Quail was heard in five different places on Lord Lilford's

Bank Hall estate, viz., in several parts of Bretherton, and on Tarleton Moss; and in the spring of 1886 it was heard on the latter, where it was certainly resident. It has been killed in each winter month, and during the hard winter 1880–1 birds were found dead in the moss-pits, to which places they had gone for food and shelter."]

[Mr. Frank Nicholson, writing in 1887, on the birds of the vicinity of Manchester, says that a few broods of Quail are reared annually on some of the fields between Withington and Chorlton, but were generally overlooked because people were unacquainted with the call-note.—Ed.

FAMILY TETRAONIDÆ.—GENUS LAGOPUS.

RED GROUSE.

Lagopus scoticus (Latham).

Local Name—*Moorgame*.

The Red Grouse is a well-known resident on all the fells and moorlands, becoming very numerous in Bowland, and the other more secluded districts. On the ling-covered mosses, both inland and adjoining the coast, it is also found the year through, but in smaller numbers than formerly, and some mosses have been deserted altogether. Although a few pairs remain to breed in these low lands, the birds are commonest in November, when a partial migration from the hills takes place; and the late Rev. J. D. Banister, in his journal, dated September 11th, 1838, remarks on their propensity at this season for oats and clover seed. In hard weather they are often met with low down in the

valleys elsewhere. The Red Grouse does not take
kindly to confinement, but Montagu records an instance
in which a pair bred in the aviary at Knowsley. He
says ("Orn. Dictionary, Supp.," 1813), "Lord Stanley
assures us that a pair of Grous which had been confined two years, by a person who paid little attention to
them, had produced many eggs. This circumstance
made his Lordship desirous to obtain the birds, in which
he succeeded, and that last year (1811) the female laid
ten eggs, which she incubated, and brought out eight
young. These infant birds, from some unknown cause,
probably a defect of natural food at that tender age, did
not live many days. The old birds feed on grain and
oatmeal, like others of the gallinaceous tribe. They
are still remarkably shy, and are as little disturbed as
possible, in order to induce them to breed again. . .
A mottled brown and white variety, very much resembling the summer plumage of the Ptarmigan, was
shot in Lancashire, in the month of August (Lord
Stanley)." Dr. Skaife, who (*Mag. Nat. Hist.*, 1837)
remarks on a male Red Grouse, which was kept in a
state of domestication by a gamekeeper of Mr. Joseph
Feilden, of Witton House, near Blackburn, for six years,
fresh ling being supplied for its use every day or two,
also notes some curious variations of plumage in four
specimens which he examined. He writes (*Mag. Nat.
Hist.*, 1838), "The first was of a pure cream colour
throughout, without spot or shade; the ground colour
of the second was of the same dusky hue, but the bird
was freckled and marked throughout with spots and
streaks of light brown; the other two birds had the
usual plumage of the Grouse, except that the wings
were white. These birds were all shot out of the same
covey or pack that season, on the moorlands east of

Blackburn." The eggs vary from six to sixteen in number, mostly ten or eleven, and are usually laid the end of March or beginning of April, but many instances of earlier dates are on record, and the Rev. W. B. Daniel ("Rural Sports," 1801-13) writes that "on March 5th, 1794, the gamekeeper of Mr. Lister (now Lord Ribblesdale) of Gisburne Park, discovered on the manor of Twitten,* near Pendle Hill, a brood of Red Grouse, seemingly about ten days old, and which could fly about as many yards at a time : this was an occurrence never known to have happened before so early in the year." A good supply of water is of vital necessity for the preservation of this species in health, and in a dry spring birds may be found lying dead in all directions on moors not thus well-provided.

GENUS TETRAO.

BLACK GROUSE.

TETRAO TETRIX, Linnæus.

LOCAL NAMES—*Black-cock, Grey-hen.*

Dr. Leigh ("Nat. Hist. Lanc., &c.," 1700) writes :— "Of Moor Game we have great plenty, both of the small and the Black, they live upon heath, but more particularly upon that part which we call Erica or Dwarf-Cypress." The present species, however, seems to have been unknown in the county for a long time until a few pairs established themselves in Furness nearly forty years ago. In his "Fauna of Lakeland," the Rev. H. A. Macpherson adduces evidence that these were artificially

* Probably a misprint for Twiston.

introduced.—Ed. Their coming was thus described by Mr. W. Pearson in the *Zoologist* for 1850, p. 2968: "We have had, within these few years, an extensive immigration of that noble bird, the Black Grouse, of which, I believe, there is no record in memory that they ever existed here before. About September, 1845, were discovered six or eight Black Game about one and a half miles from High Wray, and two were shot. They came of their own accord. Other birds have followed the larch plantations, as the Crossbill, which has been pretty numerous in Henry Curwen's woods. Black Game are found at Cock Hag, betwixt Crook and Underbarrow; at Lamb How in Crosthwaite; on the summit of Whitbarrow; on the heights of Cartmel Fell; and in the woods of Furness-fells as far down as Holker Hall. It is remarkable that, within the period of my memory, the summit of Cartmel Fell, then a heathy waste, was tenanted by the Common Red Grouse: it is now a larch forest, and occupied by Black Game." This is a most interesting example of the way in which changes in the character of a district will bring about changes in its avi-fauna, and it is gratifying to learn, as I do from Mr. Rawdon B. Lee of Kendal (*in epist.*, November 22, 1882), that the Black Grouse appears to be commoner now than when Mr. Pearson wrote, being more plentiful in the Winster district than elsewhere in the neighbourhood. It is not found in any other locality in the county, and an attempt which, Mr. T. Altham tells me, was made about the year 1864 to introduce the species at Whitewell, in Bowland, by sitting a lot of eggs, resulted in failure, the birds gradually diminishing, and disappearing altogether after two or three years.

ORDER FULICARIÆ.

FAMILY RALLIDÆ.—GENUS RALLUS.

WATER-RAIL.

RALLUS AQUATICUS, Linnæus.

LOCAL NAME—*Scarraygrise* (i.e., scared-in-the-grass,—a shy, timorous thing).

A resident species, much rarer than formerly, but probably now and at all times commoner than supposed, from the extremely retiring nature of its habits. It is very generally distributed, and has been suspected, on good grounds, of breeding in several places, but no positive evidence is in existence that I can find, though birds have many times been observed in spring and summer. In some parts of the low districts on each side of the Ribble estuary it is common in autumn and winter, and elsewhere more or less so, according as sluices, and other slow-flowing streams in marshy ground, abound or not. The late Mr. H. Durnford wrote (*Zool.*, 1874) that the stomachs of two he examined on the 12th of January contained "a little fibrous vegetable matter, remains of small mollusks, and legs of a water-beetle, with a few pieces of gravel and chalk." In the collection of Mr. W. Fitzherbert Brockholes of Claughton Hall, is a curiously pied variety, killed at the Maynes, near Poulton-le-Fylde, about forty years ago.

GENUS PORZANA.

SPOTTED CRAKE.

PORZANA MARUETTA (Leach).

The Spotted Crake is best known as an autumn and winter visitor, and although rare on the whole, there are few districts in which, at one time or other, specimens have not been shot, like the Water-Rail, its retiring habits making it easily to be passed over. No actual nest is on record, but Mr. J. B. Hodgkinson tells me that the species used to be very common on a marsh near Preston, now drained, and that he has often seen young birds there which must have been bred close by. Dr. Skaife wrote in the *Mag. of Nat. Hist.*, 1838, that he was informed by some farmers that in the neighbourhood of Martin Mere it was as abundant as the Water-Rail, and Lord Lilford writes me that near Tarleton it is not uncommon in September. Mr. Hugh P. Hornby has seen and shot a considerable number near St. Michael's-on-Wyre, between September and December, and, writing in the *Zoologist* (December, 1873) on one young and one old bird which he caught alive, says:—
" They show great repugnance to flying, preferring to trust to their legs, running very quickly and low, and looking more like rats than birds. Even when liberated in open ground, the two I caught refused to fly, though quite free from injury. On being placed in some shallow, clear water, they immediately dive, staying below the surface a considerable time, occasionally using their wings until a rushy patch be found, in which they creep, and remain as long as possible, when they raise their

heads, but no more, out of the water." The Spotted Crake is rarer in Furness; Mr. Anthony Mason of Grange tells me that he has only once seen it, when two birds were killed a few years ago by flying against the telegraph wires in that neighbourhood.

BAILLON'S CRAKE.

Porzana bailloni (Vieillot).

Mr. James Holland of Middleton states that a specimen was killed near that town by a telegraph wire in the autumn of 1870. [In May 1886 Mr. Woodburn obtained one in the clay-pits near Conishead and Ulverston, in the Furness district. See Rev. H. A. Macpherson's "Fauna of Lakeland."—Ed.]

LITTLE CRAKE.

Porzana parva (Scopoli).

A very rare visitor; the following are on record :— A specimen caught alive in a drain in Ardwick meadows, near Manchester, in the autumn of 1807, now in the Manchester Museum (*Mag. Nat. Hist.*, 1829, p. 275, John Blackwall). Has been shot on the pond at Drinkwater Park, 1860 (John Plant, *Mss.*). "North Meols" (*Naturalist*, 1837, P. Rylands). "One specimen from Crosby—Mather" (Byerley, "Fauna of Liverpool," 1856). One killed at Bradshaw Fold, 1864 (R. Davenport, *Mss.*). [In April 1886, the Mr. Woodburn above mentioned, took a Little Crake alive by means of his retriever in the clay-pits. This success led to further search, and to the capture of the Baillon's Crake. See "Fauna of Lakeland."—Ed.]

GENUS CREX.

CORN-CRAKE or LAND-RAIL.

Crex pratensis, Bechstein.

Local Names—*Daker-hen, Draker-hen, Draken-hen.*

A summer visitor; stated to arrive in the south of the county before the end of April, but does not make its appearance in most places until the early part of May. It is not usually seen after the first few days of October; but many instances are on record of its occurrence during the winter months, and there seems little doubt that occasionally individuals remain until the following spring. Various explanations of the names by which it is known among the country people have been attempted, but my own belief is that they are taken from its note, and that they simply mean the "hen" which says "drake-drake"; the fact also that about Blackburn it is called the "Draken-hen" or "Draking-hen" confirms, I think, this idea, and to those who know the northern way of pronouncing *dr* with the hard *th* sound of the Keltic *d*, it will seem as reasonable to interpret the sound to be "drake-drake" as "crake-crake." The Corn-Crake is a common bird, but is decreasing probably from the same cause as the Quail, numbers of its eggs being destroyed by the mowing-machines. Apart from special causes of this sort, however, great variations take place in its numbers in different seasons, and Dr. Skaife (*Mag. Nat. Hist.*, 1838) remarks that, whilst plentiful usually, it was rare in the year in which he was writing. Mr. W. A. Durnford (*Zool.*, 1878, 1880) says that in Furness it was scarce in 1877, and unusually abundant in 1878, and near Clitheroe it only recovered its position in 1880, after having been very infrequent for several seasons. Since 1871 Mr. Hugh P. Hornby says (*Zool.*, 1873) it has been rare at St. Michael's-on-Wyre, formerly having been a common species. In earlier times it appears to have been numerous, for Dr. Leigh writes (" Nat. Hist. Lanc., &c.," 1700), " The Rale is a bird about the bigness of a Partridge, and is common in these parts; it hides itself in the grass, and is discovered by the snarling noise that it continually makes." The Corn-Crake has often been observed to sit upon a hedge, and emit its peculiar cry, especially in the evening; but the Rev. T. Dent of Grindleton, in a letter to Dr. Garstang of Clitheroe, dated May 13, 1846, described the proceedings of one, which, a few days before, in brilliant sunshine, had perched during the forenoon, several times, for ten or twelve minutes at once, on a quickset hedge close to the house-windows, and occasionally called out as contentedly as if among the grass. It will take to the water when necessary, and the late Mr. Thomas Garnett of Clitheroe wrote in the *Field* of January 15, 1859, as follows:—" In a meadow here, on the bank of the river, which had been mowed a short time, a man was raking the cut grass from the bank, and started a Land-rail from some coltsfoot which grew there; she merely flitted across the river, and then, instead of concealing herself, as Land-rails usually do, she sat on the opposite bank making a very peculiar noise, and one I had never heard from a Land-rail before, and I watched her with some interest, when immediately about ten young ones (which from their size and appearance could not have been hatched more than a day or two) leapt from the bank into the river (which was here about thirty yards wide), and swam across with as much ease and promptitude as if they were so many ducks." The Corn-Crake lays nine or ten eggs in June.

GENUS GALLINULA.

MOOR-HEN.

GALLINULA CHLOROPUS (Linnæus).

Local Names—*Water-hen, Coot.*

Resident and common, found breeding everywhere in weedy pits and ponds, and frequenting them throughout the winter; when these are frozen, it takes to the running streams. The eggs are six or eight in number, and are laid the end of April or beginning of May; sometimes, when the nest is left for a while, they are carefully covered up with reeds. The Moor-hen is very pugnacious when it has young, and will drive away every other bird or animal which may then approach its vicinity. Mr. W. Fitzherbert Brockholes writes me that in October 1884, he had brought to him, alive and quite undamaged, a specimen of the "hairy" variety of this bird (Cf. *Trans. Norfolk and Norwich Nat. Soc.,* vol. iii. pp. 581-587, J. H. Gurney, jun.).

GENUS FULICA.

COOT.

FULICA ATRA, Linnæus.

The Coot is a resident species, but is very local, and whilst breeding numerously in many localities, in others

is never seen except as a straggler in winter. It is
more common in the southern parts of the county than
elsewhere, and near Croston and Rufford is very
plentiful, numbers being visible to travellers between
Preston and Burscough Junctions, on the pools by the
side of the railway. It breeds at Knowsley, in Claughton, Wrightington, and Woodfold Parks, and before the
drainage of Martin Mere was completed, Mr. R. J.
Howard says it nested there in hundreds. It is found
at Marton Mere, near Poulton-le-Fylde, in some numbers, and Dr. St. Clair writes me that on May 17th,
1879, he saw several nests there with eggs. In Furness
also it breeds, and Mr. W. A. Durnford says (*Zool.*,
1880, p. 241) that in the hard winter of 1879, it was
the most numerous of the wildfowl on Windermere.
[It breeds plentifully on Esthwaite Water.—R. J. H.]
The Coot lays about seven eggs some time in May,
and incubation begins before the whole are deposited.

ORDER ALECTORIDES.

FAMILY GRUIDÆ.—GENUS GRUS.

CRANE.

GRUS COMMUNIS, Bechstein.

In the *Zoologist* for 1865, Mr. C. W. Devis says that a good specimen was killed in May (exact date not ascertained) in the neighbourhood of Stretford. I have not been able to gather any confirmation of this, but a clear case has been made out for the admission of the bird to the Lancashire list by Mr. R. Milne-Redhead, F.L.S., of Bolton-by-Bowland. This gentleman, on August 25th, 1884, about 4 P.M., saw—with the naked eye and also through a glass—two Cranes flying as from W.N.W. to S.S.E.; and he states that having often observed them in Germany and other places, he is very familiar with the appearance of the species. The weather was noted by him as brilliantly fine and clear, air cooler, bar. 29·55, there having been a prevalence of easterly winds for some time before. [Cf. R. J. Howard, *Zool.*, 1884, pp. 470–471.—Ed.]

FAMILY OTIDIDÆ.—GENUS OTIS.

LITTLE BUSTARD.

OTIS TETRAX, Linnæus.

An immature male, killed at Heaton about 1840, is now in the possession of Mr. W. Fitzherbert Brockholes of Claughton Hall. Mr. Peter Rylands (*Naturalist*, 1837) gives Burscough as a locality, and the late Rev. J. D. Banister (*Mss.*) Overton, near Lancaster, but in neither case are any particulars mentioned. Mr. J. H. Swainson informs me that a specimen was shot at Thornton, near Fleetwood, on the 15th of September, 1858.

ORDER LIMICOLÆ.

FAMILY GLAREOLIDÆ.—GENUS GLAREOLA.

COLLARED PRATINCOLE.

GLAREOLA PRATINCOLA (Linnæus).

Bullock wrote (*Trans. Linnean Soc.*, xi. p. 177): "The first instance of this bird having been killed in Britain occurred in 1807, when one was shot in the neighbourhood of Ormskirk, in Lancashire; it was preserved by Mr. J. Sherlock, of that place, from whom I purchased it a few days afterwards." This specimen passed into the collection of the then Earl of Derby, and is still in the Derby Museum at Liverpool. Montagu ("Orn. Dic.") apparently referring to the same bird, says he was assured by Bullock it was shot on the 18th of May, 1804, in the act of taking beetles on wing, the remains of which were found in its stomach, and that it was thought to be a male.

FAMILY CHARADRIIDÆ.—GENUS CHARADRIUS.

GOLDEN PLOVER.

CHARADRIUS PLUVIALIS, Linnæus.

LOCAL NAME—*Yellow Plover*.

This handsome species breeds not uncommonly on many of the moorlands, and in winter is sometimes very numerous on the sea-shore. A few pairs may be found each summer frequenting almost all the hills on the Yorkshire border, from Blackstone Edge to Pendle, and on some parts of the Bleasdale and adjoining fells it is comparatively abundant. Mr. John Weld says that it appears in flocks of 20 or 30 in the low lands about Leagram Hall the middle or end of March, and that by the middle of April these leave for the neighbouring moors. It used, indeed, according to the *Mss.* of the late Rev. J. D. Banister, to breed on Pilling Moss, but those notes were written some forty years ago, and it certainly does not do so now. Inland it frequents the valleys in hard weather, flocking with Lapwings, and with them appearing on the coast by thousands should the frosts continue. These large flocks consist mostly, of course, of migrants from the north, and Mr. T. Jackson writes me that in 1881, on the 3rd, 4th, and 5th of March, immense numbers (and of Wigeon also) were packed on the shores of the Lune ready for departure, and waiting for the end of a terrific gale which was then blowing from the south-east. In autumn Golden Plover do not usually come in any numbers until October. The eggs are four in number, and are laid about the 1st of May; a nest I found on Pendle Hill on the 10th May, 1879, consisted of a rather deep and neatly-rounded hollow, the bottom being covered with about half a handful of dry bents; the position being a rather bare, grassy place, several yards from any heather, and with a good look-out over the neighbouring ground. The old bird flew away, with just one whistle when she had got about twenty yards from the nest, and did not re-appear, though I heard her whistling in the distance for nearly half an hour; even then being very shy, and flying a long way off when I moved towards

P

her. The black breast characteristic of the summer plumage is generally assumed in greater or less perfection before the shore is abandoned in spring.

GENUS SQUATAROLA.

GREY PLOVER.

SQUATAROLA HELVETICA (Linnæus).

A winter visitor, which appears on the coast at the autumn migration, remaining till spring. It is not at all uncommon, but does not gather in such large flocks as the Golden Plover; a few usually keeping together, and not associating with other waders. Not being of a very suspicious habit, its numbers are rapidly thinned on arrival by the gunners, but, generally, it is considered much scarcer now than it was some score of years ago. It has only occasionally been found inland.

GENUS ÆGIALITIS.

RINGED PLOVER.

ÆGIALITIS HIATICULA (Linnæus).

LOCAL NAMES—*Sand-lark, Tullot, Whistling Tullot, Ring-neck Purre, Pew William, Tew William, Grundling.*

The Ringed Plover is the commonest of the resident shore-birds on the Lancashire coast; breeding on the whole length of the sea-beach, and in the lower portion of all the river-estuaries. It is rapidly, however, becoming scarcer in the neighbourhood of the more frequented watering-places, and is subjected to so much disturbance from the movements of parties of visitors, that its extinction in those localities cannot be far distant. I am not aware of its nesting anywhere inland, but in winter it is frequently seen, and sometimes in considerable numbers, on many of the fresh-water reservoirs. At this season it is much more plentiful everywhere, and the flocks are, no doubt, greatly augmented by migrants. The earliest eggs are laid by the middle of April, but great numbers of birds remain unpaired a month later, and as young in down may be found up to the middle of August, it is very probable that more than one family is brought up by each pair. The shingle, just above high-water mark, is most frequently chosen as the nesting-place, though sheltered positions among the sand-hills are also favoured, and a slight hollow is scooped out, sometimes lined with fragments of shells and stones. The number of eggs is, I believe, invariably four.

GENUS EUDROMIAS.

DOTTEREL.

EUDROMIAS MORINELLUS (Linnæus).

The only case I have met with of the Dotterel's occurrence during the autumn migration happened September 22nd 1884, when five birds, going S.W., passed close by a man named Cookson, who at the time was working his cymbal-nets on Tarleton Moss. In spring, though much less plentiful than formerly, it still appears and remains for a few weeks, as it did in Pennant's time ("Brit. Zool.," 1776-77), on the fallows bordering the lower reaches of the Ribble. It occurs sometimes at Formby and the mouth of the Alt, and used regularly to frequent Pilling and Winmarleigh Mosses, being so numerous forty years ago that Mr. J. B. Hodgkinson tells me hundreds were offered for sale in Preston market in May. On the hills at the head of Wyresdale it is also seen at this season, and on Beaton Fell, on the 1st of August, 1879, a female, in good condition, and with the hatching spot bare, was shot by Mr. ——— Bates of Liverpool. The late Rev. J. D. Banister, too, writes (*Mss.*) that he has never found a nest, but that on the highest of these hills he has known a female bird to be killed with eggs in her at maturity. It is now never seen on Pendle Hill, where on August 28th 1834, a Dotterel was killed which passed into the possession of the late Dr. Garstang ; or on Blackstone Edge, where, twenty-six years ago, Mr. John Drake says it occurred regularly. Mr. T. Jackson writes me that four birds out of a flock of eighteen were shot at Pilling in 1880 on March 12th, but this is a much earlier date than usual, and from about the 26th of April to the 12th of May is the average time of its sojourn.

GENUS VANELLUS.

SOCIABLE PLOVER.

Vanellus gregarius (Pallas).

[The following appeared in *The Zoologist* for 1888, p. 389 :—" In the ' Birds of Lancashire ' (p. 175), I stated that I had examined a Cream-coloured Courser which had been shot in this county, and it is with mingled grief and joy that I have to advise you, firstly, of an error in identification, and, secondly, of an addition to the British list. The bird in question, having come into the possession of Mr. W. H. Doeg, of Manchester, was submitted by him to Mr. Seebohm, and he has pronounced it to be the Sociable Lapwing, *Vanellus gregarius* (Pall.). With the twenty years' reputation of the specimen as a Courser, and being only able to look at it by candle-light in a crowded case, perhaps I may be forgiven for having been deceived by the superficial resemblance between the two, and for having attributed such discrepancies as I could perceive, to the vagaries

of country bird-stuffing. However, the credit of the determination belongs entirely to the gentleman named; and possibly Mr. Seebohm, to whom I have sent all the information I could get, may refer to the matter more fully.—F. S. Mitchell." This specimen was exhibited by Mr. Seebohm at a meeting of the Zoological Society of London, on November 20th 1888. Cf. "Illustrated Manual of British Birds," p. 537.—Ed.

LAPWING.

VANELLUS VULGARIS, Bechstein.

LOCAL NAMES—*Pewit, Tewit, Green Plover.*

Resident, and everywhere abundant, being probably one of the commonest birds in the county. Generally, too, it is increasing, and only in such districts as that of Clitheroe, for instance, where there is very much less ploughing than there used to be, can it be stated as fewer in numbers than in former years. In winter it collects in considerable flocks, and these move about from one place to another in search of food, hard weather driving them to the sea-coast, whence they fly backwards and forwards as the severity of the season varies. The Lapwing is more a bird of the low grounds than the moorlands, and the high fells, from which spring the streams of the Brennand and Whitendale valleys in Bowland, are the only localities where I have seen them in numbers at any altitude. It is an early breeder, hatches two broods in the season, and the bulk lay their first three or four eggs in April, but every year many nests may be found in March, and in 1883 a confiding pair near Clitheroe had made all preparations, and got one egg safely deposited, on the first of that month; the storm which came a week later, however, upset their calculations, and made all the birds in the neighbourhood flock together again for shelter. The earliest nest I have heard of was reported to the *Field* of March 4th 1882, by Mr. H. J. Parke, who says that "on February 22nd, a Plover's nest was found in Brindle, near Preston, containing three eggs, and on the 25th the fourth egg was laid, when the bird commenced to sit." This bird will take to the water on occasion, swimming buoyantly, and Baron von Hügel relates in the *Zoologist* of 1872, how three young, not many days old, on being disturbed by him near the reservoir at Stonyhurst, dashed boldly in, screaming loudly, and swam about twelve yards from the bank, where they were joined by the parent bird. The Lapwing shows great attachment to its nest, defending it boldly, and being very reluctant to leave it, as the following instance will show:—About half-past five in the evening of May 15th, 1879, Mr. T. Altham found a nest with four eggs in, three of which were completely covered with a dry cake of cow-dung, probably kicked over it by accident by the cattle. The birds had evidently been trying to remove this, but had not been able. The eggs were cold, but he took them home, put them on the oven all night, and at six next morning took them to the nest again. The old birds were about the place, and the hen, on his leaving, went on at once, three of the eggs the morning following being hatched and the young gone: the remaining egg had been accidentally cracked.

GENUS STREPSILAS.

TURNSTONE.

STREPSILAS INTERPRES (Linnæus).

Young birds of this species are occasionally seen in September and October, but in spring it occurs much more commonly, and small flocks in full summer plumage frequent the islands of Walney and Foulney, the mouth of the Wyre, and the more rocky portions of the shores of Morecambe Bay, regularly during the month of May, a few individuals almost always remaining till towards the end of June. On the Ribble it used to be common, and is still taken there in small numbers, but altogether it occurs much less frequently than it once did. It is rare inland, and is only now and then seen there in stormy weather.

GENUS HÆMATOPUS.

OYSTER-CATCHER.

HÆMATOPUS OSTRALEGUS, Linnæus.

LOCAL NAME—*Sea-pie*.

The Oyster-Catcher is resident, and whilst most abundant on Walney Island, and round the shores of Morecambe Bay, is still found on the Ribble, and breeds in decreasing quantities on the sand-hills of Southport and Formby. It is very common in winter on the sand and mud-banks, and large additions accrue through the arrival of migrants, considerable flocks remaining until late in spring. The eggs are laid from early in May to mid-June, and although the nest is generally only a slight hollow scooped in the sand, it is sometimes beautifully lined with fragments of shell, and often with pieces of wood and sea-weed, and bits of straws. From being placed among the large stones and drift just above high-water mark, the eggs, which so resemble their surroundings in appearance, are not easy to find, and until the young are hatched, the birds are very wary. When this happens, however, they fly round the intruder with piercing shrieks, feigning lameness, and using every artifice to lead him away to a distance. The nest is often placed too among the bays of the sand-hills, and the eggs are three or four in number, though Mr. Howard Saunders (*Zool.*, 1866) expresses great doubts as to any single bird laying more than three. The Oyster-Catcher rarely appears inland, but is sometimes shot on the large reservoirs. Dr. Leigh remarks (" Nat. Hist. Lanc., &c.," 1700), " The Sea-pyes are very common," and there is no doubt the Lancashire coast is pre-eminently suitable to the requirements of this notable and brilliant bird.

FAMILY SCOLOPACIDÆ.—GENUS RECURVIROSTRA.

AVOCET.

RECURVIROSTRA AVOCETTA, Linnæus.

Mr. W. A. Durnford (" Birds of Walney," 1883) says that the Avocet has occurred on the shores of Morecambe Bay, and Mr. J. B. Hodgkinson informs me that a specimen passed into his possession, which was shot on the Ribble in 1840, he believes in the autumn.

GENUS PHALAROPUS.

RED-NECKED PHALAROPE.

PHALAROPUS HYPERBOREUS (Linnæus).

A rare visitor, on passage; it is said to have been shot on the Ribble.

GREY PHALAROPE.

Phalaropus fulicarius (Linnæus).

Seen on the coast when migrating in spring and autumn, and occasionally also inland. The late Dr. Garstang of Clitheroe had a specimen, shot on the Ribble near that town, in 1837, about Christmas.

GENUS SCOLOPAX.

WOODCOCK.

Scolopax rusticula, Linnæus.

It is now a well-ascertained fact that the Woodcock breeds regularly in many parts of North Lancashire, but it is best known as a migrant in autumn; appearing towards the end of October or beginning of November, and being then everywhere generally dispersed in suitably wooded localities. It remains throughout the winter, and the greater part leaves in March or April, the remainder then setting about nesting operations. Mr. W. A. Durnford (*Zool.*, 1876) says it is very plentiful in the thick coppices to the north of Barrow, where, young birds having been many times seen, he fancies it breeds in some numbers, and Mr. E. T. Baldwin (*Zool.*, 1878) does not consider its nesting in the Furness district generally at all exceptional. It has been known to rear its young in Cartmel, and near Silverdale; and, coming further south, Mr. R. Standen tells me he has an egg taken in Bleasdale Forest in 1872, where also Mr. Louis H. Simpson has known it often to breed.

Nests have been found during the last few years at Whitewell in Bowland, at Balderstone on the Ribble, and young birds only just able to fly have been seen the end of May at Langho, at Sabden, and on the Yorkshire side of the Hodder in Bashall Eaves. Woodcocks are much less common now in winter than formerly, and not many birds are caught in the 'springes' which Pennant describes ("Tour in Scotland," 1774, p. 32) as being numerous in the northern parts of Lancashire, and in which he says "multitudes are taken in this manner in the open weather," the springes being "laid between tufts of heath, with avenues of small stones on each side, to direct these foolish birds into the snares, for they will not hop over the pebbles." Woodcocks mostly frequent the thick woods which clothe the lower fells, but a continuance of hard weather will bring them down to the open brooks in the neighbourhood of the rivers.

GENUS GALLINAGO.

GREAT SNIPE.

GALLINAGO MAJOR (J. F. Gmelin).

Of the Great Snipe Montagu says ("Orn. Dic.," 1802) that it was first described as a British bird by Pennant from a specimen shot in Lancashire, and preserved in the museum of Sir Ashton Lever (Cf. "Brit. Zool.," 1776-77). Since then (the greater portion of the examples killed being young birds of the year) it has been regularly noticed all over the kingdom, but nowhere, probably, more frequently than in the county where it was first discovered, for I find over twenty specimens recorded as obtained there within about the same number of years. All these, however, have been in the autumn or early winter, and no single occurrence is reported after the turn of the year. Near St. Michael's-on-Wyre Mr. Hugh P. Hornby has been exceptionally fortunate in securing specimens, and he remarks (*Zool.*, December, 1873) on one shot September 23, 1873, that it lay very close, and on rising uttered a note not unlike that of the Common Snipe, but that it had a totally different flight, heavy, slow, and straight away. He writes me that in 1883, one, which was a perfect mass of fat and grease, was shot in the same locality in September, and a second a day or two later near Kirkham, and this last I have had an opportunity of examining.

COMMON SNIPE.

GALLINAGO CELESTIS (Frenzel).

LOCAL NAMES—*Full Snipe, Lady Snipe.*

The Common Snipe is an abundant species, breeding numerously in many localities; while in winter, with large added flocks of migrants, it is plentiful wherever there are suitable feeding-grounds. In the southern half of the county it only nests in small numbers, but as a visitor it is said to begin arriving on the mosses about the end of August, continuing to do so through September. At this time it is very wild, keeping together in "whisps" of thirty or forty, and when one gets up, and gives its alarm-note, whisps from all parts of the mosses rise up also (*Field*, November 24, 1855; *Zool.*, 1850, p. 2772, J. M. Jones). In the valleys of the Ribble and Hodder, and on the lower parts of the moors bounding them—for its nest is seldom seen so high as 1,000 feet above the sea—it breeds commonly, as is also the case on the higher reaches of the Wyre. It is a well-known resident, too, in Furness. The late Rev. J. D. Banister wrote in his journal that in the neighbourhood of Pilling the migrating flocks arrive in great numbers in October, but do not remain long, and appear to go southwards; and in 1872 Mr. Hugh P. Hornby says (*Zool.*, March, 1873) that more Snipes appeared at St. Michael's-on-Wyre than had been known for many years, and so early as the third week in July; about 260 being shot before the flocks took their departure at the end of October. Snipe are always most numerous in wet seasons, and when frost sets in the marshes are deserted for the small brooks and ditches. This species is double-brooded, and is an early breeder; the first lot of four eggs being laid the last week of March or early in April, usually on a slight elevation in damp situations, and in the least possible nest. Mr. T. Altham tells me that he once put three Lapwing's eggs into a Snipe's nest, substituting them for its own, and that they were duly hatched, but the young were promptly deserted. The Snipe is one of the earliest risers in the morning, and may often be heard drumming before day-break, but this is more generally indulged in towards evening, and in fine weather so early in the year as January. Its once discredited habit of occasionally perching on trees has now been noticed by many observers. Of the varieties of this species, Mr. Hugh P. Hornby (*Zool.*, December, 1873) records having shot "a redder-plumaged specimen than usual," and several examples of the melanic form, once accorded specific honours as Sabine's Snipe (*Scolopax sabinii*,

Vigors), have been procured at various times (*Proc. Liverpool Lit. and Phil. Soc.*, 1863-64, C. Collingwood), (*Mag. Nat. Hist.*, 1838, J. Skaife), (*Zool.*, 1879, C. S. Gregson).

SNIPE-PANTLE."

" Snipe in winter are often caught on the south side of the Ribble in the snares locally called "pantles," probably from the Anglo-Norman "panter = a net or snare" (Halliwell, "Archaic Dic.," 1850). They are formed of twisted horsehair, the main line or "rudge" being twelve yards in length, and twenty hairs in thickness, and into this the nooses, of two hairs only, and known as "guelders," are woven in pairs, about three inches apart. The "rudge" is stretched three inches from the ground, and is fastened to four pegs called "nebs," fourteen inches long, one at each end, and the other two dividing it into three equal lengths or "bows." Putting "guelders" in order for the first time is called "eyeing," and setting them after they have been used is "tilling." Snipe and Teal are mostly caught during the night, and in preparing the ground the fowlers shuffle along sideways, with the feet close together, trampling a strip of grass about a foot in width, so that in the darkness it has some resemblance to a narrow plash of water. Sky-Larks, too, are largely taken in pantles, the rudge

GENUS LIMNOCRYPTES.

JACK SNIPE.

LIMNOCRYPTES GALLINULA (Linnæus).

LOCAL NAME—*Jack Snipe.*

Although far from being as plentiful as the Common Snipe, this winter visitor is very evenly distributed, and occasionally, as Lord Lilford tells me is the case at Tarleton, appears in great numbers. Blackwall ("Researches in Zoology," 1834, p. 7) gives October 1st and April 3rd as the average dates for its arrival and departure between 1814 and 1828 in the neighbourhood of Manchester, and I have these confirmed from many other sources; but it is very often noticed by the third week in September, and Mr. Hugh P. Hornby tells me that in 1878 he saw one on the 21st of August. Generally, it is reported as occurring less frequently than formerly, but near Overton Mr. T. Jackson thinks it increases, feeding on the rushy marshes in wet, and by the streams in frosty, weather. It is a solitary species, and does not occur in flocks like the Common Snipe. Odd birds have been known to remain until very late in spring, and a correspondent of the *Field* of May 16th, 1874, says that in Prince's Park, Liverpool, he saw on

being of string, and the guelders only one hair in thickness. After a fall of snow is the most favourable time, a long narrow strip of ground being swept bare, and some wheat thrown down. Not many other birds fall victims in this way, but Lord Lilford writes (*Zool.*, 1883, p. 495), "On October 25th I received from one of our gamekeepers a very fine old male Scoter, *Œdemia nigra*, minus one foot, with a note stating that the bird was 'caught,' probably in a 'pantle' or snipe-snare, on the mere in Tarleton, on 22nd inst."

May 9th a Jack Snipe get up within two yards of him, and fly away across the lake, alighting on the further bank.

GENUS TRINGA.

DUNLIN.

TRINGA ALPINA, Linnæus.

LOCAL NAMES—*Purre, Sea-mouse.*

From the sand-banks and mud-flats of the Duddon to those of the Mersey, the Dunlin is by far the commonest of the winter shore-birds, and frequents by thousands the whole of the coast-line at that season. Not many of these stay to nest in the county, but although the mosses, &c., near the coast are almost entirely deserted now, yet on the higher hills many broods are reared each year. In the *Mag. of Nat. Hist.* for 1834, Mr. Henry Berry, referring to Bewick's account of the Purre, says that on Martin Mere he has shot the female with eggs fully matured, and found several nests from which he had driven the old birds, and Mr. R. J. Howard tells me that John Cookson, an old wildfowler, has often taken the nest in the same place. On other of the coast mosses it has also bred many times, though never very plentifully, and the *Report of the Bury Nat. Hist. Soc.*, 1871, states that it has been known to do so on Chat Moss, near Manchester. According to the *Mss.* notes of the late Rev. J. D. Banister, it bred on the hills at the head of Wyresdale, some thirty years ago, and R. Leyland, of Halifax, says (*Mag. Nat. Hist.*, 1828) that it "breeds on Blackstone Edge." It is still found nesting on the more northern parts of the latter range of moors,

and from near Todmorden I have myself examined a
clutch of young in down, and have been informed that
Dunlin appear there every summer. On May 9th, 1880,
I saw a fine black-breasted bird there, but could not
raise its mate, and newly-hatched young were found in
1876 on May 20th, and in 1877 on May 28th. Mr. T.
Altham has seen it on the Bowland fells, and on Pendle
it has been observed several times in full breeding-
plumage, but a thoroughly-authenticated nest has never,
to my knowledge, been taken there. In 1876, on May
20th, Mr. W. Purnell found three eggs, which, unfortunately, had been disturbed, and one sucked, and their
connection with several Dunlin which were seen on the
hill at the same time could not be traced. In Furness,
Mr. W. A. Durnford states that it is resident, but does
not nest, and small flocks, indeed, may be seen in every
month of the year along the whole of the shore. By
the end of April the winter flocks are broken up, the
black breast of the summer plumage has been assumed,
and the birds leave for their breeding-quarters, returning towards the end of July, and gradually increasing
until the enormous quantities which are seen in October
are again collected. Mr. T. Jackson says he has shot a
large quantity at a time by moonlight at high water,
and that they seem to collect on the edge of the sea in
the night-time; they are by no means shy during the
day, and are always easy of approach. The Dunlin is
not often seen inland in winter, but Mr. R. Davenport
tells me that it has many times been shot on the Bury
reservoir and at other places in the neighbourhood.

LITTLE STINT.

Tringa minuta, Leisler.

The Little Stint occurs on migration in spring and autumn, but is by no means common. At both seasons it has been noticed inland, and Mr. R. Davenport informs me that on the 14th September 1870, four specimens were shot on the Bury reservoir, two of which remain in his possession. The same gentleman saw a pair on Whittle Pike in May, about the year 1875, and on the moors east of Burnley it has for many years been observed in that month, in suitable, damp weather, and even so late as June. Those familiar with the ground say that "a pair or two of birds resembling Dunlins, but without the black breast and not much bigger than Larks, are seen nearly every spring, and that they have always been spoken of as Little Dunlin." They are described as being very familiar in their habits, running like mice among the tufts, squeaking, and permitting approach within five or six yards.

TEMMINCK'S STINT.

Tringa temmincki, Leisler.

A rare visitor in spring and autumn. Mr. J. E. Harting writes me that in 1864 one was shot on Ribbleton Moor by Mr. Sharples, from whom he purchased it in May of that year; and the Rev. E. D. Banister presented a specimen to the Lancaster Museum, which he shot at Pilling in 1873, but in what month he cannot remember.

CURLEW-SANDPIPER.

Tringa subarquata (Güldenstädt).

An uncommon visitor on passage, having been shot in autumn from the end of August to the beginning of October, and in spring from April to the end of May, by which time the summer plumage has been generally assumed.

PURPLE SANDPIPER.

Tringa striata, Linnæus.

The sandy shores of Lancashire are not suitable for this rock-loving species, and although during the winter months it is a regular visitor in small numbers, and has been shot along the whole line of coast, it is one of the rarest of the Sandpipers. It has been killed inland, but very infrequently.

KNOT.

Tringa canutus, Linnæus.

Local Names—*Dunn, School-bird* (pronounced *Scoo-bird*).

The Knot is a well-known spring and autumn migrant, and, especially on Walney Island and round Morecambe Bay, flocks of thousands may be seen at those seasons, as also throughout the winter if the weather be hard. Willughby ("Ornithology," Ray, 1678, p. 302) says, "In the month of February in the year of our Lord 1671, on the coast of Lancashire about Leverpool I observed many of this sort of birds flying in company." It is not, however, so common on the Mersey and Ribble as in the localities before mentioned, where it gets its local name through such large quantities keeping so close together. It is rarely that birds are taken with any remnant of the brilliant colours of the summer plumage, but Dr. Kershaw of Middleton has a specimen—shot at Southport in September, 1883, and which he saw in the flesh—with a decidedly red breast. Inland it is very occasional.

GENUS MACHETES.

RUFF.

MACHETES PUGNAX (Linnæus).

A spring and autumn migrant, occurring regularly in small numbers, but now being much more uncommon than some twenty or thirty years ago. It used then to be taken in large quantities on the Ribble (see p. 171), and although Pennant ("Brit. Zool.," 1776–77) says that, visiting Martin Mere the latter end of March or beginning of April, it did not continue there above three weeks, there appears little doubt he was in error, and that it remained to breed. Mr. R. J. Howard states (*Zool.*, 1884, p. 466) that there are mere-men still living who can remember the birds remaining all the summer, have seen them assume and throw off the "ruff," and have often watched them at the "hill"; while he has two males in full breeding-plumage, which were shot on Martin Mere about 1840, and near the same time a young bird, unable to fly, was caught by William Parker of Crossens. The Ruff has occurred several times inland in autumn, and Mr. H. P. Hornby writes me that in September, 1881, he shot a male and two females out of a small flock near St. Michael's-on-Wyre. At this season, young birds and Reeves (females) are most frequently met with.

GENUS CALIDRIS.

SANDERLING.

CALIDRIS ARENARIA (Linnæus).

The Sanderling is a regular migrant in spring and autumn, and individuals have occasionally been shot in the winter months. Unlike its congeners, it appears to frequent in greater numbers the mouth of the Mersey than the more northern parts of the shore, and Mr. H. Durnford (*Zool.*, July, 1873) speaks of flocks in the spring of that year exceeding those of the Dunlin: he also remarks on the well-known preference of the species for drier situations, and sand-banks rather than mud-flats, and probably this is in part the reason why the one locality is more favoured than the others. It remains very late in the spring, and Mr. James Cooper, writing in 1845 (*Zool.*, p. 1192), says he shot it on June 6th on the banks of the Ribble, and has seen it even later: such birds have always assumed the complete breeding-plumage. It does not appear to be noticed in autumn on the coast much before September, and inland is not at all uncommon, being sometimes, as Mr. R. Davenport tells me, plentiful on the Bury reservoir.

GENUS TRYNGITES.

BUFF-BREASTED SANDPIPER.

TRYNGITES RUFESCENS (Vieillot).

A male bird was killed at Formby, on the banks of the river Alt, about thirteen miles north of Liverpool, in May, 1829, and was sent to Liverpool market for sale along with some Snipes. This specimen passed into the possession of the Rev. T. Staniforth, late of Bolton Rectory, Skipton, who transmitted the record to Yarrell ("Brit. Birds," 4th ed., vol. iii. p. 436).

GENUS TRINGOÏDES.

COMMON SANDPIPER.

TRINGOÏDES HYPOLEUCUS (Linnæus).

LOCAL NAMES—*Summer-Snipe*, *Sand-Snipe*, *Sand-pie*, *Sand-lark*, *Dicky-di-dee*, *Willy-wicket*.

A summer visitor, arriving about the middle of April, and leaving in September. The Common Sandpiper prefers the neighbourhood of fast-running, gravelly streams, and in the south of the county and on the lower reaches of all the rivers it is seldom seen except on migration. In the hilly districts, however, it is more or less common everywhere, and in some localities exceedingly so. Mr. H. Kerr says that in Rossendale it is fairly numerous in the breeding-season, on the Calder and Ribble it is common, while on the Hodder and its tributaries in Bowland it is especially plentiful.

In Higher Wyresdale it is well known, and round the tarns and lakes of the Furness fells it breeds regularly, as also on all the streams which flow into Morecambe Bay. The eggs are four in number, and are laid about the beginning of May, in a slightly-constructed nest, usually a very little way from the water, among the short docks and herbage. Sometimes, however, it is built as much as one hundred yards off (*Zool.*, 1872, A. von Hügel), and an early nest taken by Mr. T. Altham on April 25th, 1875, and which had four fresh eggs, was close to a footpath quite forty yards from a stream. The young are able to run about immediately on being hatched, and the late Mr. Thomas Garnett of Clitheroe described (*Mag. Nat. Hist.*, 1833, p. 148) an instance, in which attempts at escape were made by chicks which had only been out of the shell an hour or two. Both young and old readily take to the water, and the latter are expert swimmers and divers, able, even if winged, to baffle the utmost efforts of a dog to seize them.

SPOTTED SANDPIPER.

Tringoïdes macularius (Linnæus).

In a paper on the Notabilia of the Archæology and Natural History of the Mersey District during the years 1863, 1864, 1865, by Mr. H. Ecroyd Smith, published in the *Proceedings of the Historic Society of Lancashire and Cheshire*, Session 1865-66, is a note on this North American species by Mr. C. S. Gregson, who says: " Edwin Lord of Warrington shot two on the Mersey below that town in May, 1863, one of which I possess.

In 1865 he again saw this species on the river, but did not get a shot." Mr. Gregson has written me, under date June 12, 1884, that he saw in all four specimens, in the flesh, and in process of skinning by Lord. [Cf. J. H. Gurney's "Rambles of a Naturalist," pp. 255-262, and Yarrell's "Brit. Birds," 4th ed., iii. p. 453.—Ed.

GENUS HELODROMAS.

GREEN SANDPIPER.

HELODROMAS OCHROPUS (Linnæus).

The Green Sandpiper is one of the earliest to arrive of the autumn migrants, and a few birds may be seen each year in one place or another, frequenting in August the fresh-water brooks and runnels, and regularly occurring from that month until December. I do not find any record of examples killed in the spring, but the species has been shot many times in July, and the late Dr. Skaife (*Mag. Nat. Hist.*, 1837) was confident that a male and female killed near the river Darwen on July 29th and August 1st respectively in 1837, and one or both of which had been seen about for two or three weeks, had been breeding: this, however, is extremely doubtful, as no authentic nest has ever yet been taken in Britain. The Green Sandpiper is essentially a feeder by fresh water, and even when near the sea is always found on the banks of ponds, &c., at some distance from the shore.

GENUS TOTANUS.

WOOD-SANDPIPER.

TOTANUS GLAREOLA (J. F. Gmelin).

A rare visitor, and, curiously enough, all the occurrences I have heard of, except one recorded by Mr. N. Cooke, of Warrington (*Zool.*, Sept. 28, 1848, p. 2304), are from the Calder and Ribble in the neighbourhood of Mytton and Whalley. Mr. W. Naylor preserved one shot at Calder foot on November 18, 1867, and has a second killed in May, 1869, in Mytton wood, which had a mate with it, and the late Mr. David Mitchell, about four years before, saw two old birds near the same place, having four young with them, which he believed to have been bred there, for he averred that they were only able to fly just sufficiently well to avoid capture, and owing to this the parents were quite close to him many times. Mr. H. Miller also tells me that he has mounted two birds which were shot near Cock Bridge, on the banks of the Calder—one in 1879, in August, he thinks. Mr. Cooke says of his specimen that it was shot as it rose from a pit, in company with some Snipes, and that it was identified by the late Mr. H. Doubleday.

REDSHANK.

TOTANUS CALIDRIS (Linnæus).

This species is common on the coast from August to April, and occasional stragglers appear in winter inland. The great majority, however, leave in spring, and although the Redshank is never scarce during the breeding months, its nest has seldom been found within the county limits. Mr. J. B. Hodgkinson has seen eggs on Walney, which had been taken there, and is confident that it breeds in the Winster valley, but the Cumberland coast is the nearest point where it nests in any numbers. Mr. T. Jackson says that it is common all the year round on the Lune, frequenting the creeks on the marshes, and that he considers it to be largely on the increase.

SPOTTED REDSHANK.

Totanus fuscus (Linnæus).

A rare visitor, occurring irregularly on passage. Mr. C. S. Gregson (*Proc. Histor. Soc. Lanc. and Chesh.*, 1865-66) says he shot one at the mouth of the Alt in October, 1864, and Mr. H. Miller writes me, that about 1873 he had a specimen which was killed on the edge of the Calder above Whalley. Mr. P. J. Hornby shot a male bird on August 22nd, 1877, near St. Michael's-on-Wyre, and Mr. J. B. Hodgkinson says that a pair in summer plumage were seen on the Ribble in May, many years ago, by the late James Cooper. In October, 1883, about the 17th, as Mr. W. Fitzherbert Brockholes informs me, one was sent him shot near Pilling.

In *The Zoologist*, 1889, p. 109, Mr. Chas. F. Archibald, of Rusland Hall, Ulverston, reports upon a specimen in sooty plumage, shot in the month of April a few years previously.—Ed.

GREENSHANK.

TOTANUS CANESCENS (J. F. Gmelin).

The Greenshank is a regular visitor on migration in autumn, and has been shot as late as November, but it only appears in small numbers, and in spring is still more rarely met with. It is often seen in company with Redshanks.

GENUS MACRORHAMPHUS.

RED-BREASTED SNIPE.

MACRORHAMPHUS GRISEUS (J. F. Gmelin).

An example of this rare American species was obtained near Southport in 1873, about September, and passed into the collection of Mr. J. B. Hodgkinson, by whom it was presented to the Preston Museum. ["In September, 1891, Mr. Hodgkinson saw a Red-breasted Snipe hanging in the shop of Maudsley, game dealer, Preston, and learned that it had been purchased from a man who sold it as a Common Snipe. He bought the bird, had it stuffed by Gillett, Lancaster Road, Preston, and kindly lent it to me."—R. J. H.]

GENUS LIMOSA.

BAR-TAILED GODWIT.

LIMOSA LAPPONICA (Linnæus).

A spring and autumn migrant, seldom appearing at the latter season until September, or remaining longer than the end of the following month. According to the late Rev. J. D. Banister (*Mss.*) the return flight commences in March, but Mr. W. A. Durnford says (*Zool.*, 1876), that in 1876 he saw a flock of Godwits, probably Bar-tailed, on February 19th, on Walney, and in the same locality, in 1881, Mr. Hugh P. Hornby informs me he noticed several small parties as late as the 4th of June. Mr. J. B. Hodgkinson still considers it a common autumn visitor to the Ribble estuary, but so long ago as 1838, the late Dr. Skaife, in the *Mag. of Nat. Hist.* of that year, wrote of it as only being thinly scattered along the shore, and refers to the immense flocks he used to meet with when shooting there.

BLACK-TAILED GODWIT.

LIMOSA BELGICA (J. F. Gmelin).

The Black-tailed Godwit is not as common as the Bar-tailed, but is a regular visitor in autumn, and Mr. T. Jackson writes me that on the Lune he sees both species there in small numbers, all having left, however, by the end of October. Near St. Michael's-on-Wyre, Mr. Hugh P. Hornby says they have twice shot specimens : one, a male, on September 23, 1873, and the second (three birds having been seen the previous day) on September 12, 1882. Mr. C. S. Gregson has only met with this species once on the Formby shore. I do not find any record of its occurrence on the spring migration.

GENUS NUMENIUS.

WHIMBREL.

NUMENIUS PHÆOPUS (Linnæus).

LOCAL NAMES — *May-bird, Curlew-hilp, Curlew-whilp, Curlew-whelp.*

The "Curley-hilps" mentioned by Dr. Leigh as sometimes being on "the smaller Martin Meer in great numbers" (see p. 140), no doubt mean Whimbrels, and the name, with slight variations, is still used on the Ribble and Wyre, and most probably is to be interpreted as a little Curlew, or a "whelp" off a Curlew. Young Whimbrels are seen on passage in September and October, and occasionally a few birds may remain the winter; but at both seasons the species is rare, and it is during the vernal migration only that it becomes common, considerable flocks appearing along the coast in May, and the regularity of their visits providing it with another of its local names. The Whimbrel sometimes alights on the moors in autumn, and a correspondent of the *Field* of September 2, 1871, says that in the previous week he shot a female out of two birds which he came across on Stansfield Moor, near Todmorden; in September 1883, also, Mr. Jon. Dean got one on the hills above Rossendale. In the spring it frequents the pastures near the coast on its first arrival, afterwards gravitating towards the sand-banks and the sea-edge, and Mr. C. S. Gregson (*Nat. Scrap Book*, pt. 4) has known it to remain very late on some of the mosses, though he is certainly in error in thinking it sometimes breeds.

CURLEW.

Numenius arquata (Linnæus).

The Curlew is a common resident, breeding more or less plentifully on all the fells, and frequenting the coast in winter in large numbers, where also many birds may be seen the year round. Its nest is found pretty regularly on some of the low-lying mosses, and on Chat Moss it occasionally breeds (*Report Bury Nat. Hist. Soc.*, 1871), whilst on those of Cockerham, Foulshaw, &c., it is a constant summer resident. In the south of the county it is only observed on passage, and as far north as Burnley it is rarely seen even on the higher lands, for, as Mr. H. Kerr remarks on the Rossendale district, the neighbourhood of the manufacturing towns, and the fact of the moors being open to all, and intersected by footpaths, are quite sufficient reasons for the scarcity of all shy and wary species like the present. On Pendle Hill there are always a few pairs, and it breeds commonly on all the fells which form the gathering ground of the Hodder, the Wyre, and the Lune, becoming especially numerous on those at the head of Croasdale. Mr. T. Altham says, too, that it appeared in 1880 on Longridge Fell, and that in 1882 there were three pairs breeding there. On the Furness Fells it is common, and many birds here doubtless feed on the coast, but as a rule the Curlew does not travel far when on its breeding-ground, and in winter it remains entirely by the sea, not passing to and fro according to the weather, as is the habit of the Lapwing. The migratory movement begins in April, earlier if it be a mild spring, and the flocks may be heard

RING OR FLY NETS.

passing over at night in what must be very large numbers. By July the moors are again deserted, and old and young travel down to the shore together. The eggs are three or four in number, and are laid the end of April or beginning of May, in a slightly-constructed nest, sometimes on the driest parts of the ground, and sometimes in marshy places, and are rarely found except by accident. "As wary as a Curlew" has almost become a proverb, and both birds-nester and wild-fowler know this well, but Mr. T. Jackson says that in foggy weather, or in the half-light of morning or evening, it is the easiest of birds to get a shot at, and in the dark is the worst seeing one he knows. It varies very much in size, and specimens are often procured very little bigger than a Whimbrel.

"Fly-nets" or "Ring-nets" (see Plate) are set when there is no moon, and across the banks which are last covered by the tide. They are made of very fine cotton or linen thread, from three-inch to five-inch mesh, and thirteen to fifteen mesh deep, and are hung diamond-ways on poles from ten to twenty yards apart; this is done very loosely, so as to give plenty of "bag," and on the Ribble the bottom is allowed to come close to the ground, but on the Lune a gap of three feet is left. The name "Ring-nets" arises from there being a small brass ring or pulley at the top of the poles through which the cord for pulling them up is run. There is, of course, no limit to their length, and as much as 800 yards has been known to be set by one man. Curlew, Whimbrel, Geese, Ducks, and all the shore-birds are taken, the smaller ones, such as Dunlin, getting so entangled as many times to need being torn in pieces before they can be removed. The only birds which ever break through are Teal, a bunch of these, flying down wind, often doing so.

ORDER GAVIÆ.

FAMILY LARIDÆ.—SUBFAMILY STERNINÆ.
GENUS STERNA.

ARCTIC TERN.

STERNA MACRURA, Naumann.

LOCAL NAMES—*Sparling, Kek, Skrike, Sea-Swallow.*

The Arctic Tern is a summer visitor, and breeds in considerable numbers on Walney, and some of the surrounding islets. Various opinions are held as to whether it or the Common Tern is most plentiful. Mr. Howard Saunders, who visited Walney early in June, 1865, writes me that *then* they were about equal, but that near Piel there was a preponderance of Arctic; the late Mr. H. Durnford, however, as I am informed by his brother, Mr. W. A. Durnford, considered, some ten years later, the Common Tern to be the more numerous, and the latter gentleman says that three out of four of his skins are of this species, while he believes the proportion is in reality still greater.* There is no doubt that proportions change, and Mr. Saunders says that on the Farne islands, for instance, the Common Tern is distinctly pressing back the Arctic, and in Brittany also is driving away the Roseate (Yarrell, "Brit. Birds," 4th ed., iii. p. 545). The Arctic Tern arrives on its breeding-ground in May, and leaves in September or

* On May 30th, 1885, during a short visit to Walney, I recognized plenty of Common Tern, but not one Arctic, and the same at Ravenglass.—Ed.

October, being seen on migration in the other parts of the county, and occasionally being driven inland by storms. The eggs are two or three in number, and are laid about the beginning of June, often in slight depressions scratched in the sand or shingle, though sometimes a few bents are collected together, and the favourite situations are either the hollows of the sandhills, or along the shore just outside. The late Mr. John Hancock (*Trans. Nat. Hist. Soc. Northumb. and Durham*, vol. vi. p. 142) says, under Roseate Tern (though he has since written to me, in December, 1882, that he is now more inclined to think the specimen was a young Arctic Tern):—"While on an ornithological tour to the west coast (on July 27th, 1840) my attention was arrested by a Tern on the sands at Morecambe Bay; it was making the most extraordinary movements, and was evidently in trouble; so intent was it on rubbing its head from side to side upon the sand, that it allowed me to approach within gun-shot. I killed the bird, and to my surprise found a cockle firmly fixed on the upper mandible, which was inserted nearly half an inch between the valves of the shell, and was indented by its grasp: a rather strange example of the biter bit."

COMMON TERN.

STERNA FLUVIATILIS, Naumann.

LOCAL NAMES—*Sparling, Kek, Skrike, Sea-Swallow.*

The present species and the one last described are not easily to be distinguished on the wing, except by a practised eye, and at pretty close quarters, and there is little doubt that mistakes in identification are often made. Their habits, times of arrival and departure, and the conditions under which they are seen away from the shore, are much the same, but the Common Tern is the more generally distributed of the two, and is, or has been until very lately, found nesting in several localities other than Walney Island and its vicinity. On Foulney, in 1840, Mr. John Hancock found it plentiful, and Mr. W. A. Durnford ("Birds of Walney," 1883) says that it nests in suitable places all along the coast of Furness. Mr. John Watson has known it to breed for at least three years on the mosses on the Westmorland border, on Martin Mere, according to Mr. R. J. Howard, it used to do so plentifully, and not very long ago a considerable colony frequented the sand-hills in the neighbourhood of Formby. Here, on June 11, 1873, the late Mr. H. Durnford (*Zool.*, 1873) found it breeding in numbers, the two or three eggs being placed on the top of the most naked sand-hills, and without any nest whatever. On Walney it chooses sometimes a bank of pebbles just above high-water mark, often hollows in the drift sea-weed and in the sand-hills abutting on the beach, and occasionally a nest is constructed of little bits of drift-wood, a habit which Mr. J. E. Harting says is not uncommon with this species. The eggs are laid early in June, but owing to indiscriminate robbery, young in down may be found up to the end of July, and from this cause the Common Tern is now but rare in places where not many years ago its eggs might have been collected by hats-full.

ROSEATE TERN.

STERNA DOUGALLI, Montagu.

The Roseate Tern was found breeding on the island of Foulney in 1840 by Mr. John Hancock, who writes me that, to the best of his recollection, it and the Common Tern were about equal in numbers there. His visit was paid on the 27th July, but the place had been so harried that year, that the birds had had to lay second or third clutches, and six eggs he found were just hatching, the young from one, indeed, making its appearance whilst he was watching the old bird. Since then the species has become rarer and rarer, and is now probably extinct there. Mr. J. E. Harting, who visited the locality on May 30, 1864 (*Zool.*, 1864) saw several birds, but did not shoot any, and Mr. Howard Saunders, who went specially for the Roseate Tern, early in June, 1865, only saw one pair. He writes (*Zool.*, ss. pp. 181-2) that whilst traversing the isle of Walney " suddenly a harsh ' crake ' caught my ear, and there, above our heads, easily distinguishable by their more slender form, bathed in an indescribable pink glow, hovered a pair of veritable Roseate Terns. I gazed at these objects of my search until my eyes ached, but they mounted higher and higher, and amongst the score of nests in the space of half an acre round us, it was useless to attempt to identify their eggs, and I may as well say at once that this was the only time I was able to distinguish this species with perfect certainty on Walney." I have not heard of any occurrence since Mr. W. A. Durnford wrote (*Zool.*, 1876) that a pair had been shot at Biggar, on Walney, in 1874, and stuffed by a blacksmith in Barrow, who described them as " Rose-breasted Sparlings."

LITTLE TERN.

Sterna minuta, Linnæus.

Local Names—*Sparling, Kek, Skrike, Sea-Swallow, Sea-Mouse.*

Like the Arctic and Common Terns, the Little Tern is a summer visitor to the Lancashire coast, but owing to the senseless destruction which has pursued it and many other beautiful and interesting species, it now breeds only in one part of Walney Island, and that in much diminished numbers. In the neighbourhood of Lytham, where, near the lighthouse, Mr. J. B. Hodgkinson says it used to breed thirty years ago, and between that place and Blackpool, where, as Mr. R. J. Howard tells me, the nest has been taken by Mr. W. L. Constantine, the increase of population is sufficient reason for its disappearance; but on Foulney Island, once so prolific in bird-life, there is nothing save the havoc caused by shooting parties, and this so long favoured spot is now rarely frequented by any breeding birds, save perhaps a few Ringed Plovers. The Little Tern arrives in May, and leaves in October, and at these times, and especially the latter, it is often shot on the various parts of the coast, heavy storms also driving it inland. It breeds in small colonies, laying three, and occasionally four, eggs in a slight hollow in the sand or shingle, little above high-water mark, and, unlike the Common and Arctic Terns, preferring the sea-ward side of the sand-hills to the more sheltered positions inside. Its cry also is different from theirs, being a single sharp note frequently repeated (J. E. Harting, *Zool.*, 1864).

GULL-BILLED TERN.

Sterna anglica, Montagu.

The late Dr. Skaife stated (*Mag. Nat. Hist.*, ii. p. 530, 1838) that he had a specimen shot at Blackpool in the summer of 1832.

SANDWICH TERN.

Sterna cantiaca, J. F. Gmelin.

Local Names—*Sparling, Kek-Swallow, Kek, Skrike*.

The Sandwich Tern is the last, and almost the rarest, of those of its genus which breed in Lancashire. Like them, it is only a summer visitor, and, except after severe storms, is seldom seen elsewhere than in the immediate vicinity of its breeding-ground on Walney. There, nesting as it does, close to the colony of Black-headed Gulls at the north end, it has long been strictly protected, but in spite of this it appears to decrease rather than increase, although, following the Gulls, a few pairs have of late years frequented the south end of the island. It lays earlier than the other Terns, and Mr. W. A. Durnford (*Zool.*, 1880, p. 241) saw two nests, each with two eggs in, on May 25th in that year, whilst he states (*Zool.*, 1878) that on June 21st, 1877, he came across a young one almost ready to fly. Mr. J. E. Harting thus describes (*Zool.*, 1864) a visit he paid to the colony on Walney on May 30, 1864 : " As we approached they rose perpendicularly to a great height, keeping up a succession of harsh screams, not unlike the sound produced by running a sharp stick across a comb. The nests in structure were very similar to those of the Black-headed Gulls, being composed entirely of grass, and placed quite close to each other on the ground, the only difference being that the Gulls' nests were placed on somewhat level ground, whereas those of the Sandwich Tern were situate on the side of a sand-hill among long thin grass. Standing still for a minute I counted seventeen nests, all close to each other, all containing eggs, and the majority having three . . . . . the birds were on their nests between 6 and 7 A.M."

GENUS HYDROCHELIDON.

BLACK TERN.

HYDROCHELIDON NIGRA (Linnæus).

A rare visitor, occurring on migration in spring and autumn. It is seen as frequently inland as on the coast, and Mr. R. Davenport tells me that he has four young birds which were shot September 4, 1878, on the Bury reservoir. Mr. H. Miller says that he has seen it flying about Morecambe Bay up to the end of May.

SUBFAMILY LARINÆ.—GENUS RISSA.

KITTIWAKE.

RISSA TRIDACTYLA (Linnæus).

Like the other rock-breeding Gulls the Kittiwake does not nest in the county, although it occurs the whole year round, and in May is common in Morecambe Bay.

It is frequently seen inland in spring, following the courses of the rivers, and Mr. James Garnett of Clitheroe says that it *invariably*, and in pairs, comes up the Ribble in April and May, when the young salmon are making for the sea. Stragglers, especially after stormy weather, have been seen everywhere, and in winter it is more or less numerous all along the shore, being generally in company with the Common and Black-Headed Gulls.

GENUS LARUS.

GLAUCOUS GULL.

Larus glaucus, O. Fabricius.

Occasionally seen in winter, but very rarely. The late Mr. H. Durnford stated (*Zool.*, 1874) that he picked up an immature bird, with a broken wing, on the Formby shore, on November 8, 1873.

HERRING-GULL.

Larus argentatus, J. F. Gmelin.

Local Name—*Silver-back*.

The Herring-Gull is found on the coast, and occurs in flocks, especially about Walney Island, the year through, but does not nest in the county. It is occasionally seen following the courses of the rivers, and is shot at rare intervals on the inland reservoirs. Mature birds are uncommon.

LESSER BLACK-BACKED GULL.

Larus fuscus, Linnæus.

Although now confined, in the breeding-season, to one locality, the Lesser Black-backed Gull, not many years ago, used to frequent several of the Lancashire mosses, and Mr. J. B. Hodgkinson tells me a few pairs nested regularly up to a recent date at Pilling, he having seen eggs there, whilst on Cockerham Moss, Mr. T. Jackson says, it used to breed in some numbers. It is, however, only on the low grounds round the estuary of the Kent, on the borders of Lancashire and Westmorland, that, with a few Herring-Gulls, through strict protection, it has been enabled to hold its own, and probably the larger part, if not the whole, of the area over which it is found is in the latter county. It is a well-known resident, and may be seen on the coast, especially about Morecambe Bay, at all times of the year, and, in stormy weather, frequently occurs inland. Mr. W. A. Durnford ("Birds of Walney," 1883) says that this species and the Herring-Gull feed largely on shell-fish, and in this way do a considerable amount of injury to what has always been an important industry in the district: he also writes (*Zool.*, 1877) that the Lesser Black-backed Gull is said to nest on some of the islands in Windermere, but of this I have no confirmation.

The eggs are almost invariably three in number, and are laid about the end of May.

COMMON GULL.

LARUS CANUS, Linnæus.

A regular visitor from August to April, a few birds always remaining throughout the summer also, but not breeding. Like most of the other Gulls, it is often driven inland by storms.

GREATER BLACK-BACKED GULL.

LARUS MARINUS, Linnæus.

This species is no doubt the Great Gull of which Dr. Leigh ("Nat. Hist. Lanc., &c.," 1700) speaks, and which he says is nearly as large as a Goose, and breeds in vast quantities in the isle of Walney. Here, however, it has long been absent, but on Pilling Moss it has nested within recent times, and the late Rev. J. D. Banister in his *Mss.* writes, under date August 14, 1839, "The Black-Backed Gull (*Larus marinus*) is abundant on Pilling sands and breeds on the mosses, lays two or three eggs, and the young birds are hatched in the latter part of June or beginning of July." A correspondent of the *Field* of April 30, 1859, signing J. H. S., stated that it had then been driven away from Pilling, by being shot, and by having its nests destroyed, owing to its killing the young of the Black-headed Gulls, which then bred there, and it does not now nest in any part of the county.* It may be seen throughout the year, but is commonest in winter, in immature plumage, and does not usually collect in flocks, but seeks its food in pairs. Mr. T. Jackson informs me that

---

* Mr. C. F. Archibald informs me that an isolated pair breed on the fells in Furness.—Ed.

both the Greater and Lesser Black-backed Gulls feed largely on dead fish, and the carcases of various animals washed up on the shore, and that in the stormy weather of January, 1883, there were several hundreds gorging themselves on a quantity of bacon which had been cast up on the banks of the Lune. The present species is almost unknown away from the coast, and I have no records of its occurrence anywhere except in the immediate neighbourhood of the sea.

BLACK-HEADED GULL.

Larus ridibundus, Linnæus.

Local Names—*Sea Maw*, *Patch* or *Petch*, *Gor*, *Turnock*.

The Black-headed Gull is a common resident, and frequents the coast in very large numbers from August to March, retiring in the latter month to several well-known breeding-places, where in colonies numbering hundreds, or even thousands, of birds, it devotes the remainder of the year to the rearing of its young. The north end of Walney Island appears to have been longest tenanted, and it is probably from hence that proceeded the founders of the community which once occupied the site of the town of Fleetwood. Increase of population and spread of buildings drove these last about the year 1833 to Pilling Moss, across the Wyre, and there they flourished up to 1876, when, their breeding-ground being effaced by the plough, they moved on again to Winmarleigh Moss, where, having been protected by [the late] Lord Winmarleigh and Captain Bird, the joint owners, they will no doubt remain for a long time. On Walney many offshoots from the original colony have from time to time planted themselves in various parts of the island, but owing to disturbance, none of these became permanent until within recent years; now, however, on several parts of the south end, there are small communities formed, and these are said to be increasing. An old-established, but not extensive, Gullery on a floating island on a small tarn upon the Bleasdale fells, exhausts the Lancashire list, and this appears to be becoming deserted, the birds in the present year (1884) not having returned up to the beginning of May. The few eggs which the late Mr. H. Durnford (*Zool.*, 1873) found near Formby in 1873 do not appear to have resulted in the establishment of a permanent settlement. Large flocks of Black-headed Gulls follow the plough, and feed on the larvæ turned up by it, thus doing valuable service to the farmer, and the late Rev. J. D. Banister wrote (*Zool.*, 1844, p. 577) that in the twilight they also feed largely on the night-flying moths, he having repeatedly shot birds at a late hour, with their pouches crammed with them. The same gentleman has some interesting remarks (*Zool.*, 1845, p. 881) on the change of the herbage on Pilling Moss due to the presence of the Gulls. He says " originally on this moss the common Wild Duck, Teal, Snipe, Curlew, Golden Plover, Dunlin, and even Red Grouse, bred extensively. A few pairs still occupy certain districts, where they annually breed. The Black-backed and Black-headed Gulls have within the last twelve years succeeded the ancient colonists. The poor heath, in the vicinity of their breeding-place, has been almost annihilated by their excrement, and in its place has sprung up a rich and varied vegetation. No one who formerly knew this moss, and has witnessed the recent remarkable change, doubts for a moment that it has been entirely effected by the dung of these birds, deposited on the moss during the breeding-season. For, as far as the nests have extended, and even somewhat farther, the change in the herbage may be distinctly traced." The birds first appear on the breeding ground early in March, the date varying slightly according to the state of the weather, and a continual increase goes on till towards the end of April. Eggs may be found the middle of that month, but early May is the most usual time, and nearly all the young are hatched by the beginning of June. Old and young take their departure between the end of June and the beginning of August. The eggs, which are very varied in shape and markings, are three or four in number, and are laid at first simply in a slight depression of the ground, or in a tuft of rushes, but as incubation progresses a more or less carefully built nest of dry grass, rushes, or straws, is reared round them. The Black-headed Gull is not at all uncommon in many inland localities in spring, or after stormy weather.

LITTLE GULL.

Larus minutus, Pallas.

A rare winter visitor; the records are:—

"One killed at Formby" (Byerley, "Fauna of Liverpool, 1856"). One shot near Preston, and now in the possession of Mr. Salisbury, Ashton-on-Ribble. One shot November 1st, 1880, on the Mersey, off New Brighton (W. Bell, *Zool.*, 1881, p. 27). "A specimen was brought to a Barrow bird-stuffer about five years ago" (W. A. Durnford, "Birds of Walney," 1883).

One near Rossall, January 1889, Mr. Hugh P. Hornby in letter to R. J. Howard.—Ed.

SUBFAMILY STERCORARIINÆ.—GENUS STERCORARIUS.

GREAT SKUA.

STERCORARIUS CATARRHACTES (Linnæus).

An uncommon visitor, usually occurring in autumn, but sometimes in winter, and odd birds have now and then been seen well on into the summer months. All the Skuas are called locally Sea-Hawks, and by the coast fishermen a less euphonious name is also given them, from their habit of pursuing and forcing Gulls to disgorge their food, which, when dropped, is popularly supposed to be their fæces.

POMATORHINE SKUA.

STERCORARIUS POMATORHINUS (Temminck).

A pretty regular visitor in autumn, few years passing without one or two specimens being shot in September or October. In October, 1879, when the great irruption of Skuas took place on the Yorkshire coast (*Zool.*, 1880, p. 18, T. H. Nelson), one was shot on the Ribble near Samlesbury; and after storms this species has several times been taken in other localities inland.

RICHARDSON'S SKUA.

STERCORARIUS CREPIDATUS (J. F. Gmelin).

Has been shot from October to April, but is not common, and away from the coast is very rarely seen.

A fair proportion of the birds killed are in mature plumage.

BUFFON'S OR LONG-TAILED SKUA.

STERCORARIUS PARASITICUS (Linnæus).

A rare visitor, of which I only find the following occurrences noted:—

A mature bird shot August 20, 1850, at Fleetwood (*Zool.*, 1850, p. 2925, John Plant).

One shot on the Ribble, near Preston, March, 1877 (*E. mus.* Hugh P. Hornby, *fide* J. Frankland).

Three shot on the sands near Grange, October 25, 1859, the day the *Royal Charter* was lost (A. Mason, *Mss.*).

" Two adult specimens, shot on the coast, have come under my observation" (*Mag. Nat. Hist.*, 1838, John Skaife).

About a dozen examples were obtained in October, 1891, out of a flight which visited Walney Island and Morecambe Bay.—Ed.

ORDER TUBINARES.

FAMILY PROCELLARIIDÆ.—GENUS PROCELLARIA.

STORM-PETREL.

PROCELLARIA PELAGICA, Linnæus.

A winter visitor, seldom seen except during and after heavy gales. At such times it is often driven on the shore in considerable numbers, many specimens having been taken far inland, and in storms from the east as well as the west.

LEACH'S PETREL.

PROCELLARIA LEUCORRHOA, Vieillot.

Like the last species, the Fork-tailed Petrel only occurs in stormy weather, almost always from October to December, and so many specimens have been taken as to render a detailed list unnecessary. In 1881 nearly a score of captures were recorded, these being mostly on the coast, but a large number of birds have been obtained in inland localities; such being generally either found dead, or in an extremely exhausted state.

WHITE-FACED PETREL.

Procellaria marina, Latham.

[An example of this species—also called the Frigate Petrel—was washed up dead on the outside of Walney Island, after a severe gale in November, 1890. After identification, by Mr. Osbert Salvin, F.R.S., full particulars of the occurrence were communicated by the Rev. H. A. Macpherson to *The Ibis*, 1891 (pp. 602-4), and have been repeated with slight additions in his "Fauna of Lakeland." This species breeds in the Southern hemisphere, and, like many other Petrels, wanders northwards; it is well known as a visitor to the Canary Islands.—Ed.]

WILSON'S PETREL.

Oceanites oceanicus (Kuhl).

[A bird of this species was obtained at the same place as the preceding, but a day or two earlier, and has been recorded by Mr. Macpherson in *The Ibis*, with the above.—Ed.]

GENUS PUFFINUS.

MANX SHEARWATER.

Puffinus anglorum (Temminck).

A winter visitor, irregular in its appearance, but seen occasionally off the coast, and sometimes found dead or exhausted on the shore.

GENUS FULMARUS.

FULMAR.

FULMARUS GLACIALIS (Linnæus).

Of very rare occurrence, seen now and then during winter storms.

In December, 1881, one, in the possession of Mr. Hugh P. Hornby, was shot near Fleetwood, and about the same time Mr. R. Drummond obtained another at Blackpool. A keeper in the employ of Mr. J. W. Makant, Bolton, shot a Fulmar in December, 1883, on one of the Rivington reservoirs, and on January 14th, 1884, Mr. John Wrigley (*Field*, February 9, 1884, p. 169) picked up an adult bird on the Formby shore. This last was still alive, but died after a few days, and on dissection was found to be half-choked by a bone firmly fixed in the lower part of the throat.

ORDER PYGOPODES.

FAMILY COLYMBIDÆ.—GENUS COLYMBUS.

GREAT NORTHERN DIVER.

COLYMBUS GLACIALIS, Linnæus.

A few individuals appear every winter on the coast, and specimens have been obtained from early autumn till late in April. This Diver has also frequently occurred on Windermere, and most of the other large inland lakes and reservoirs.

BLACK-THROATED DIVER.

COLYMBUS ARCTICUS, Linnæus.

A rare winter visitor on the coast. Dr. Skaife writes (*Mag. Nat. Hist.*, 1838), "In the winter of 1835-36 I saw a remarkably fine young specimen in the hands of an animal-preserver at Preston : it was captured below Lytham." Mr. T. Jackson tells me that he occasionally sees and shoots it on the Lune.

RED-THROATED DIVER.

COLYMBUS SEPTENTRIONALIS, Linnæus.

A regular winter visitor on the coast, and the commonest of the Colymbidæ. Odd birds have now and then been killed in the summer months, but Mr. H. Miller informs me that it usually leaves Morecambe Bay about the end of April, appearing again in October or November. It is not often seen on the inland lakes, and is relatively much rarer in such places than the Great Northern Diver.

FAMILY PODICIPEDIDÆ.—GENUS PODICIPES.

GREAT CRESTED GREBE.

PODICIPES CRISTATUS (Linnæus).

The Great Crested Grebe is a resident species, breeding regularly in at least one locality; but in most parts of Lancashire is only known as a casual winter visitor. When frozen out from the lakes and meres, which it frequents throughout the country in summer, it migrates to the sea, and is then often seen close in shore on the coast and river-estuaries. Probably it is when on its way thither, and reluctant to leave the fresh water except as a last resource, that it halts sometimes by the larger streams, reservoirs, and lakes (until they too become ice-bound), and on many of these it is of pretty regular occurrence. On the lake in Knowsley Park Mr. J. J. Hornby writes me that there are always two or three pairs, except when, in very hard weather, the water is frozen up, and that they breed regularly, though the Swans destroy some of the nests; on the 26th of June, 1879, he saw a nest containing two eggs.

RED-NECKED GREBE.

Podicipes griseigena (Boddaert).

A winter visitor, of uncommon occurrence. I have records of only seven examples obtained within the last twenty years, six of which were from inland positions, and the seventh from the Wyre estuary.

SLAVONIAN GREBE.

Podicipes auritus (Linnæus).

A winter visitor, rarer even than the last.

One was shot in November, 1848, on Beswick reservoir (*Zool.* [1850], p. 2924, John Plant): a young bird of the year in February, 1864, near the mouth of the Alt (C. S. Gregson, "Proc. Histor. Soc. Lanc. and Chesh.," 1865-66): Bury reservoir, 1874 (R. Davenport): Fleetwood, December, 1879 (Dr. Kershaw): near Colne, winter of 1880 (H. Whalley).

BLACK-NECKED or EARED GREBE.

Podicipes nigricollis, C. L. Brehm.

[Mr. Hugh P. Hornby, writing to Mr. Howard under date of July 22nd, 1892, says:—"I have a beautiful specimen of an Eared Grebe (*P. nigricollis*), killed by R. Bagot near Lune mouth, late in March or early in April, 1886."—Ed.]

LITTLE GREBE.

PODICIPES FLUVIATILIS (Tunstall).

LOCAL NAMES—*Douker, Foot-in-arse*.

The Little Grebe is resident, and is still found breeding on a few suitable pieces of water, whilst in winter it is very generally distributed in small numbers. It has nested on the Calder near Whalley, and Mr. H. Miller tells me that, along with the Coot and Moor-hen, it used to do so on the reservoirs attached to Messrs. Steiner's printworks at Church, but that, about the year 1867, these were dug out too deep for vegetation, and all the birds left. Near Walton-le-Dale Mr. J. B. Hodgkinson thinks it may still breed, as it certainly used to, and Mr. R. J. Howard writes me that nests are found every year near Rufford and Croston, as also in Woodfold Park. From Chamber Hall, near Bury, Mr. R. Davenport had eggs in 1876, and Mr. J. J. Hornby has found several nests on Knowsley Great Water, where it is probably a regular summer resident. Mr. W. A. Durnford ("Birds of Walney," 1883) writes that it is resident in the Furness district, but that personally he has only observed it on Windermere, and that in winter. Here it has long been well known at this season, and in Camden's " Britannia," translated from the edition of 1607, and enlarged by Richard Gough, 2nd ed., 1806, vol. iii. p. 406, it is stated that " waterfowl in great plenty resort to this lake, especially in winter; such as wild Swans, wild Geese, Ducks,

Mallard, Teal, Widgeons, Didappers,* Gravyes (which are larger than Ducks, and build in hollow trees) and many others." There are few large sheets of water in Lancashire where, in winter, the Little Grebe has not occasionally occurred, sometimes pretty numerously, and it may be that these in part are migrants, though there are never such numbers as to preclude the possibility of their all having been bred within the county limits.

FAMILY ALCIDÆ.—GENUS ALCA.
RAZORBILL.
ALCA TORDA, Linnæus.

The Razorbill is essentially a pelagic species on the Lancashire coast, and—except in heavy weather, when many are washed up dead—is only to be seen a mile or two away from the land. It is plentiful from early autumn until May, and odd birds remain throughout the summer. There are a few rocky scarps on the Furness coast not unsuitable as nesting situations, but only in one case, where Mr. H. Miller found an unfledged young one on the sands in the first week of August 1880, is there any evidence of its breeding.

GENUS LOMVIA.
COMMON GUILLEMOT.
LOMVIA TROILE (Linnæus).

The Guillemot, like the Razorbill, is seldom seen except at some distance from the shore; dead or half-

---

* Willughby, and Bewick after him, gives *didapper* as a name for the Little Grebe, but what a *gravye* is I can't imagine. Mr. J. Kirby, of Ulverston, tells me that the Red-throated Diver is often called *gravvyner* in that neighbourhood.

dead specimens being washed up after stormy weather. In the autumn and winter it is pretty common, and from February to May very large flocks congregate, in Morecambe Bay especially, but during the breeding months it is only represented by a few individuals. Sometimes, but rarely, it occurs on the inland waters in winter, and occasionally too the bridled form is obtained.

GENUS URIA.

BLACK GUILLEMOT.

URIA GRYLLE (Linnæus).

A rare winter visitor on the coast. The "Report of the Bury Natural History Society" (1871) states that a young bird was found dead at Summerseat by Mr. H. Pickup of that place.

GENUS MERGULUS.

LITTLE AUK.

MERGULUS ALLE (Linnæus).

A rare winter visitor, occasionally washed up dead on the shore, or found exhausted, both near the coast and at considerable distances inland, after heavy storms.

GENUS FRATERCULA.

PUFFIN.

FRATERCULA ARCTICA (Linnæus).

LOCAL NAMES—*Couter-neb, Old Wife*.

Occurs at sea off the coast in autumn and winter, and is often washed up dead after severe weather, at times being driven some distance inland.

INDEX.

Auk, Little, 265
Avocet, 217

" Bessie-dowker " (Dipper), 31
" Billy-biter " (Great Tit), 36
Bittern, " Bitter-bump," 146
—— American, 147
—— Little, 145
Blackbird, 7
Blackcap, 17
" Blackcap " (Great Tit), 36
Blackcock, 197
" Blue-back " (Fieldfare), 6
" Blue-bill " (Scaup), 172
" Bottle-Tit," 35
Brambling, 70
Bullfinch, 75
" Bullspink " (Chaffinch), 69
Bunting, Corn-, 77
—— " Grey," 77
—— Black-headed, 79
—— Cirl, 79
—— Lapland, 80
—— Ortolan, 79
—— Reed-, 79
—— Snow-, 81
—— Yellow, 78
Bustard, Little, 206
Buzzard, Common, 123
—— Honey-, 130
—— Rough-legged, 124

Chaffinch, 69
Chiffchaff, 22
" Chitty," " Chitty-Wren," 40

Chough, 84
Coot, 204
" Coot " (Moor-hen), 204
Cormorant, 141
Corn-Crake, 202
" Couter-neb," (Puffin), 265
Crane, 206
" Crane " (Heron), 143
Creeper, Tree-, 58
—— Wall-, 60
Crossbill, 76
Crow, Carrion-, 87
—— Hooded, " Manx," 88
Crake, Baillon's, 201
—— Corn-, 202
—— Little, 201
—— Spotted, 200
Cuckoo, 112
Curlew, 238
" Curlew-hilp " (Whimbrel), 237
Cushat, 182

" Daker-hen," " Draker-hen," 202
Decoys, 162
" Deviling " " Devil Screamer," 101
" Dicky-di-dee " (Sandpiper), 230
Dipper, 31
Diver, Black-throated, 260
—— Great Northern, 260
—— Red-throated, 260
Dotterel, 211
Dove, Ring-, 182
—— Rock-, 184
—— Stock-, 183

INDEX.

Dove, Turtle-, 185
Duck, Eider, 176
—— Ferruginous, 169
—— Gadwall, 169
—— Garganey, 169
—— Goldeneye, 175
—— Long-tailed, 176
—— Pintail, 167
—— Pochard, 173
—— Scaup, 172
—— Sheld-, 160
—— Shoveler, 170
—— Teal, 170
—— Tufted, 171
—— White-eyed, 175
—— Wigeon, 162
—— Wild, 168
Dunlin, 224
" Dunn " (Knot), 227
" Dunnock," " Dykie " (Hedge-Sparrow), 31

Eagle, Golden, 125
—— Spotted, 125
—— White-tailed, 126
Eider Duck, 176
" Eutick " (Whinchat), 12

Falcon, Greenland, 131
—— Peregrine, 131
—— Red-footed, 137
" Fell-Peggy " (Wood-Warbler), 25
" Fern-Owl," 117
Fieldfare, 6
Fire-crest, 21
" Fleckie " (Chaffinch), 69
" Flick-tail " (Stonechat), 13
" Flinch " (Goldfinch), 61
Flycatcher, Pied, 54
—— Spotted, 54
" Flying Toad " (Nightjar), 108
" Frank " (Heron), 143
Fulmar, 259

Gadwall, 169
Gallinule, Purple, xi
Gannet, 142

Garden-Warbler, 18
Garganey, 169
Glead (Kite), 128
Godwit, Bar-tailed, 235
—— Black-tailed, 236
Goldcrest, 20
Goldeneye, 175
Goldfinch, 61
" Goldfinch " (Yellow Bunting), 78
Goosander, 179
Goose, Barnacle, 154
—— Bean-, 153
—— Brent, 154
—— *Canada*, xi
—— *Egyptian*, xi
—— Grey Lag-, 150
—— Pink-footed, 153
—— White-fronted, 153
" Gor " (Black-headed Gull), 252
" Gorse-cock " (Linnet), 71
Gos-Hawk, 127
" Grass-check " (Whinchat), 12
Grasshopper-Warbler, 29
Grebe, Black-necked or Eared, 262
—— Great Crested, 261
—— Little, 263
—— Red-necked, 262
—— Slavonian, 262
Greenfinch, " Greenbull," 63
Greenshank, 235
Grey-hen, 197
Grosbeak, Pine-, 75
Grouse, Black, 197
—— Red, 195
Guillemot, Black, 265
—— Common, 264
Gull, Black-headed, 252
—— Common, 251
—— Glaucous, 249
—— Great Black-backed, 251
—— Herring-, 249
—— Kittiwake, 248
—— Lesser Black-backed, 250
—— Little, 254

Harrier, Hen-, 121
—— Marsh-, 121
—— Montagu's, 122

INDEX.

Hawfinch, 64
Hawk, Gos-, 127
—— "Night-", 103
—— " Red " (Kestrel), 137
—— Sparrow-, 127
" Hazel-Linnet " (Lesser White-throat), 16
Hedge-Sparrow, 31
Hobby, 132
Honey-Buzzard, 130
Hoopoe, 111
Heron, 143
—— Night-, 146
—— Purple, 145

Ibis, Glossy, 148

Jackdaw, 86
Jay, 85

Kestrel, 137
Kingfisher, 109
Kite, 128
—— *Swallow-tailed*, 130
Kittiwake, 248
Knot, 227

Lapwing, 214
Lark, Shore-, 100
—— Sky-, 95
—— " Tit-", 47
—— Wood-, 99
Linnet, Brown, Grey, 71
—— Green, 63
—— " Manx," 73

Magpie, 85
Mallard, 168
Martin, House-, 57
—— Sand-, 58
" May-bird " (Whimbrel), 237
Merganser, Red-breasted, 180
Merlin, 135
Mistle-Thrush, 1
Moor-hen, 204
" Mussel-cracker " (Goldeneye), 175

Night-jar, " Night-hawk," 103

Nightingale, xii
" Nope," 36, 39, 75
Nuthatch, 40

Oriole, Golden, 49
Osprey, 139
Ouzel, " Fell-", 8
—— Ring-, 8
—— " Rock-", 8
—— " Water-" (Dipper), 31
Owl, Barn-, 114
—— Little, 120
—— Long-eared, 115
—— Scops-, 120
—— Short-eared, 117
—— Tawny, or Wood-, 118
—— Tengmalm's, 119
—— White, 114
" Ox-eye " (Great Tit), 36
Oyster-catcher, 216

Partridge, 192
—— *Red-legged*, xii
Pastor, Rose-coloured, 83
" Patch " (Black-headed Gull), 252
Petrel, Fulmar, 259
—— Leach's, 257
—— Storm-, 257
—— White-faced, 258
—— Wilson's, 258
Pewit, 214
Phalarope, Grey, 218
—— Red-necked, 217
Pheasant, 192
Pigeon, " Hill-", 183
—— Wood-, 182
Pintail, 167
Pipit, Meadow-, 47
—— Richard's, 48
—— Rock-, 49
—— *Tawny*, xiii
—— Tree-, 47
—— *Water*-, xiii
Plover, Golden, 208
—— " Green " (Lapwing), 214
—— Grey, 210
—— Ringed, 210
—— Sociable, 212
Pochard, 173

INDEX.

Pratincole, Collared, 208
Puffin, 265
Purre (Dunlin), 224

Quail, 193
—— *Virginian*, xii

Rail, Land-, 202
—— Water-, 199
Raven, 92
Razorbill, 264
Redbreast, " Robin," 15
" Redcap " (Goldfinch), 61
Redpoll, Lesser, 72
—— Mealy, 73
Redshank, 233
—— Spotted, 234
Redstart, Black, 14
—— " Red-tail," 14
Redwing, 5
Ring-Ouzel, 8
Roller, 111
Rook, 91
Ruff (female, Reeve), 228
" Russiannet " (Wigeon), 162

Sanderling, 229
Sand-Grouse, Pallas's, 186
" Sand-lark," 210, 230
Sandpiper, Buff-breasted, 230
—— Common, 230
—— Curlew-, 227
—— Green, 232
—— Purple, 227
—— Spotted, 231
—— Wood-, 233
" Scarragrise " (Water-Rail), 199
" Scarth " (Cormorant), 141
Scaup, 172
" School-bird " (Knot), 227
Scoter, Common, 176
—— Surf-, 179
—— Velvet, 178
" Sea-Pheasant " (Pintail), 167
" Sea-Pie," 216
" Seed-fool " (Yellow Wagtail), 46
Shag, 141
Shearwater, Manx, 258
Sheld-Duck, 160
Shoveler, 170

Shrike, Great Grey, 50
—— Red-backed, 51
—— Woodchat, 53
Siskin, 63
" Silver-back " (Herring-Gull), 249
Skua, Buffon's, 256
—— Great, 255
—— Pomatorhine, 255
—— Richardson's, 255
Sky-Lark, 95
Smew, 181
Snipe, Common, 220
—— Great, 219
—— Jack, 223
—— Red-breasted, 235
—— " Summer," 230
" Sparling," 242, 243, 246, 247
Sparrow-Hawk, 127
Sparrow, Hedge-, 31
—— House-, 66
—— " Reed-", 79
—— Tree-, 67
Spoonbill, 148
" Spoonbill " (Shoveler), 170
Stannel (Kestrel), 137
Starling, 82
—— Rose-coloured, 83
Stint, Little, 226
—— Temminck's, 226
Stonechat, 13
Swallow, 56
Swan, Bewick's, 157
—— *Mute*, xi
—— *Polish*, xi
—— Whooper, 156
Swift, 101
—— White-bellied, 102

Teal, " Throstle-Teal," 170
Tern, Arctic, 242
—— Black, 248
—— Common, 243
—— Gull-billed, 247
—— Little, 246
—— Roseate, 245
—— Sandwich, 247

Thrush, Mistle-, 1
—— Song-, 3

INDEX.

Titlark, 47
Titmouse, Blue, 39
—— Coal-, 37
—— Great, 36
—— Long-tailed, 35
—— Marsh-, 38
"Tullot" (Ringed Plover), 210
"Turnock" (Black-headed Gull), 252
Turnstone, 215
Twite, 73
"Tweedler" (Merlin), 135

Wagtail, Grey, 45
—— Pied, 44
—— "Rock-", 45
—— White, 42
—— Yellow, 46
Warbler, Blackcap, 17
—— *Dartford*, xiii
—— Garden-, 18
—— Grasshopper-, 29
—— Reed-, 26
—— Sedge-, 27
—— Willow-, 23

Warbler, Wood-, 25
Water-hen, 204
Waxwing, 53
Wheatear, 9
—— Black-throated, 11
Whimbrel, 237
Whinchat, 12
"White Robin" (Spotted Flycatcher), 54
Whitethroat, 16
—— Lesser, 16
Wigeon, 162
Willow-Wren, 23
"Willy-wicket" (Sandpiper), 230
Windhover (Kestrel), 137
Woodchat, 53
Woodcock, 218
Woodpecker, Green, 107
—— Great Spotted, 104
—— Lesser Spotted, 106
Wood-Wren, 25
Wren, 40
Wryneck, 108

Yellow-hammer, 78

AN
Illustrated Manual of British Birds,
BY
HOWARD SAUNDERS, F.L.S., F.Z.S., &c.,
Editor of the Third and Fourth Volumes of the Fourth Edition of "Yarrell's History of British Birds,"

In One Volume, 750 pages demy 8vo, cloth, with 367 fine Woodcuts, and 3 Maps, £1 1s.

"It would be difficult to give a better condensation of facts in fewer lines than has been contrived by Mr. Saunders. . . . We have only to add, that the natural history loving portion of the British public ought to be grateful to Mr. Saunders for having placed within reach at a moderate cost, and in one volume, such a well-illustrated and accurately-written account of our native birds."—*Zoologist.*

"Perhaps the question most frequently put to a zoologist by a lay friend is, 'What is a really good book on British Birds that is not too expensive?' and the question has been one that has been found extremely difficult to answer. Mr. Saunders deserves our thanks for having taken this difficulty out of our way."—*Athenæum.*

"Excellent alike in style and matter, it ought to be in the hands of every lover of birds, and should take the place of several inferior books on the subject now before the public."—*Annals of Natural History.*

"It is scarcely necessary to inform those who are acquainted with the previous work of the author, that the information is not only valuable from its correctness, but that it is brought up to the present date."—*Field.*

LONDON:
GURNEY & JACKSON, 1, PATERNOSTER ROW
(SUCCESSORS TO MR. VAN VOORST).

THE FOWLER IN IRELAND:

OR,

NOTES ON THE HAUNTS AND HABITS OF WILD FOWL AND SEA FOWL,

INCLUDING

Instructions in the Art of Shooting and Capturing Them.

BY

SIR RALPH PAYNE-GALLWEY, BART.

In One Volume, 8vo, 504 pages, £1 1s.

With many Illustrations of Fowling Experiences, Birds, Boats, Guns, and Implements, drawn by the AUTHOR *and Mr.* C. WHYMPER.

"No more suitable book for a country house can be imagined during the long winter evenings."—*Academy.*

"More particular than 'Folkard's Wildfowler,' and free from the antiquated details of Colonel Hawker's book, this treatise cannot fail to be of service to the sportsman. The first half of this book is so valuable we have not lingered long over the technical details of the other half. Every lover of birds will enjoy the first half of this book."—*Athenæum.*

THE BOOK OF DUCK DECOYS:

Their Construction, Management, and History.

BY

SIR RALPH PAYNE-GALLWEY, BART.

Crown 4to, cloth, 226 pages, with Coloured Plates, Plans, and Woodcuts, £1 5s.

"We believe that in regard to Antiquarianism, Natural History, or Sporting (and it has something in common with all), it is many years since so original or so curious a volume has appeared."—*Spectator.*

"In this handsome volume with its coloured plates and diagrams there is much interesting and even romantic reading. Sir Ralph Payne-Gallwey is an enthusiast in sport, and especially in all manner of wild-fowling."—*Times.*

LONDON:

GURNEY & JACKSON, 1, PATERNOSTER ROW

(SUCCESSORS TO MR. VAN VOORST).

Bird-Life of the Borders:

RECORDS OF
WILD SPORT AND NATURAL HISTORY
ON MOORLAND AND SEA.

BY
ABEL CHAPMAN.

Demy 8vo, 300 pages, with 50 Illustrations by the Author, 12s. 6d.

"At last we have a book on birds in their haunts by a writer who is thoroughly master of his subject—one who has plenty to say, and who also knows how to place his experiences vividly before the reader. The portions devoted to the Cheviots and the moorlands recall the scent of the heather, while the narrative of adventures by day and by night in a gunning punt along the 'slakes' off Holy Island is pervaded by the keen salt breezes from the North Sea. . . . . As regards the second part, which treats of wild-fowling with the stancheon-gun, we can only say that nothing like it has appeared since the publication of Colonel Hawker's classic work. The haunts and habits of wild-fowl by day and night have never before been so clearly pointed out in any work with which we are acquainted."—*Athenæum.*

"One of the pleasantest books conceivable. . . . Every lover of a country life will delight in his vivid sketches of sporting experience and wild life on the moors. . . . The author's enthusiasm is something irresistible. Even the drawbacks of that 'waiting game,' wild-fowling appear as of no weight when estimating the glories of the sport as set forth in the admirable chapters on 'Wild-Fowl of the North-East Coast,' 'Midnight on the Oozes,' 'Wild-Fowl and the Weather,' and so forth."
—*Saturday Review.*

"An invigorating out-of-door air pervades this book, and a happy directness of description. . . . Although very comprehensively treating of bird-life, a considerable portion of the book—and that not the least interesting—is devoted to shooting (open and covert), but mainly punt shooting. In sporting experience, so far as concerns the north-east coast, Mr. Chapman stands in the front rank, and discourses of it with an authority beyond controversy or challenge."—*Land and Water.*

"His pages bristle with curiously minute and interesting facts concerning 'our feathered friends.'"—*Leeds Mercury.*

"Will enchant all who are fond of birds. Sympathy with all living creatures, careful observation with cautious deductions, and strong love for the bleak moors and wild scenery of the Cheviots—such are the characteristics of this most interesting book. . . . The illustrations add a great charm to a book redolent of wild life and careful observation."—*Academy.*

LONDON:
GURNEY & JACKSON, 1, PATERNOSTER ROW
(SUCCESSORS TO MR. VAN VOORST).

Researches in Zoology,

ILLUSTRATIVE OF THE

STRUCTURE, HABITS, AND ECONOMY OF ANIMALS.

By JOHN BLACKWALL, F.L.S.

Second Edition, 350 pages.   8vo, 7s. 6d.

CONTAINING

Chapters on the Migration of Birds; The Notes of Birds; Observations on the Cuckoo; Remarks on the Swallow Tribe; On the supposed capability of Periodical Birds to become Torpid; On the Instincts of Birds; Observations on the Pied Flycatcher; Remarks on Bewick's Swan; Malformation of the bill of Birds; On the Nudity of the Head of the Rook; The Diving of Aquatic Birds; The Grenadier Grosbeak, &c., &c. On the Growth of the Salmon and Sewin, and Notes on Spiders.

8vo, sewed, price 6d.

A List of British Birds,

REVISED TO JULY, 1892, BY

Howard Saunders, F.L.S., F.Z.S., &c.

(SECOND THOUSAND.)

For Labelling Specimens or for Reference.

LONDON:
GURNEY & JACKSON, 1, PATERNOSTER ROW
(SUCCESSORS TO MR. VAN VOORST).

www.ingramcontent.com/pod-product-compliance
Lightning Source LLC
Chambersburg PA
CBHW021959220426
43663CB00007B/880